CHARACTERIZATION OF TRIBOLOGICAL MATERIALS

MATERIALS CHARACTERIZATION SERIES

Surfaces, Interfaces, Thin Films

Series Editors: C. Richard Brundle and Charles A. Evans, Jr.

Series Titles

Encyclopedia of Materials Characterization, C. Richard Brundle, Charles A. Evans, Jr., and Shaun Wilson

Characterization of Metals and Alloys, Paul H. Holloway and P.N. Vaidyanathan

Characterization of Ceramics, Ronald E. Loehman

Characterization of Polymers, Ned J. Chou, Stephen P. Kowalczyk, Ravi Saraf, and Ho-Ming Tong

Characterization in Silicon Processing, Yale Strausser

Characterization in Compound Semiconductor Processing, Yale Strausser

Characterization of Integrated Circuit Packaging Materials, Thomas M. Moore and Robert G. McKenna

Characterization of Catalytic Materials, Israel E. Wachs

Characterization of Composite Materials, Hatsuo Ishida

Characterization of Optical Materials, Gregory J. Exarhos

Characterization of Tribological Materials, William A. Glaeser

Characterization of Organic Thin Films, Abraham Ulman

CHARACTERIZATION OF TRIBOLOGICAL MATERIALS

EDITOR

William A. Glaeser

MANAGING EDITOR

Lee E. Fitzpatrick

BUTTERWORTH-HEINEMANN
Boston London Oxford Singapore Sydney Toronto Wellington

MANNING
Greenwich

This book was acquired, developed, and produced by Manning Publications Co.

Design: Christopher Simon
Copyediting: Margaret Marynowski
Typesetting: Stephen Adams

Library of Congress Cataloging–in–Publication Data
Characterization of tribological materials / editor, William A. Glaeser : managing editor, Lee E. Fitzpatrick.
p. cm.—(Materials characterization series)
Includes bibliographical references and index.
ISBN 0–7506-9297-9 (acid-free paper)
1. Tribology.   2. Materials—Mechanical properties.   I. Glaeser, William A.   II. Series.
TJ1075.C514   1993
621.8'9—dc20   93-24421
                                                                                          CIP

Butterworth-Heinemann
80 Montvale Avenue
Stoneham, MA 02180

Manning Publications Co.
3 Lewis Street
Greenwich, CT 06830

10 9 8 7 6 5 4 3 2 1

Printed in the United States of America

Contents

Preface to Series

This *Materials Characterization Series* attempts to address the needs of the practical material user, with an emphasis on the newer areas of surface, interface, and thin film microcharacterization. The series is composed of a leading volume, *Encyclopedia of Materials Characterization*, and a set of about 10 subsequent volumes concentrating on characterization of individual material classes.

In the Encyclopedia, 50 brief articles (each 10–18 pages in length) are presented in a standard format designed for ease of reader access, with straightforward technique descriptions and examples of their practical usage. In addition to the articles there are one-page summaries for every technique, introductory summaries to groupings of related techniques, a complete glossary of acronyms, and a tabular comparison of the major features for all 50 techniques.

The 10 volumes in the Series on characterization of particular materials classes include volumes on silicon processing, metals and alloys, catalytic materials, integrated circuit packaging, etc. Characterization is approached from the materials user point of view. Thus, in general, the format is based on properties, processing steps, materials classification, etc., rather than on a technique. The emphasis of all volumes is on surfaces, interfaces, and thin films, but the emphasis varies depending on the relative importance of these areas for the material class concerned. An appendix in each volume reproduces the relevant one-page summaries from the Encyclopedia and provides longer summaries for any techniques referred to that are not covered in the Encyclopedia.

The concept for the series came from discussion with Marjan Bace of Manning Publications Company. A gap exists between the way materials characterization is often presented and the needs of a large segment of the audience—the materials user, process engineer, manager, or student. In our experience, when, at the end of talks or courses on analytical techniques, a question is asked on how a particular material (or processing) characterization problem can be addressed the answer often is that the speaker is "an expert on the technique, not the materials aspects, and does not have experience with that particular situation." This series is an attempt to bridge this gap by approaching characterization problems from the materials user's end rather than from that of an analytical technique expert.

We would like to thank Marjan Bace for putting forward the original concept, Shaun Wilson of Charles Evans and Associates and Yale Strausser of Surface Science Laboratories for help in further defining the series, and the editors of all the individual volumes for their efforts to produce practical, materials use based volumes.

C.R. Brundle   C.A. Evans

Preface

Characterization of Tribological Materials was written to illustrate the ways in which surface characterization is being used in tribology and the expected future trends. Since tribology is a discipline involving the moving contacts of surfaces, it should not be surprising that surface science must play a role. Although materials used in bearings, gears, sliding seals, brakes, clutches, electrical contacts, and magnetic recording devices have been developed expressly for these applications, the materials are not unique. Most have been adapted from conventional engineering materials. For tribological use, however, parts require some surface characterization. Currently, surface analysis during manufacture includes the determination of roughness, optical properties, surface hardness, and surface coating thickness and bond strength. More sophisticated surface analysis is not, as a rule, used routinely—except for magnetic recording media. Advanced surface analytical techniques are used mostly in the investigation of the mechanisms of friction, lubrication, and wear.

This volume presents several chapters which describe the basics of tribological phenomena and examples of the use of characterization equipment to further the understanding of these phenomena. These chapters also serve to show where surface science can play a role in advancing our knowledge of friction, wear, and lubrication.

Two chapters describe current uses of advanced surface characterization techniques for routine inspection of manufactured components (rolling contact bearings and magnetic recording media).

William A. Glaeser

Acronyms

ADAM	Angular Distribution Auger Microscopy
AED	Auger Electron Diffraction
AES	Auger Electron Spectroscopy
AEM	Analytical Electron Microscopy
AFM	Atomic Force Microscopy
ATR	Attenuated Total Reflection
BSE	Backscattered Electron
CARS	Coherent Anti-Stokes Raman Scattering
CBED	Convergent Beam Electron Diffraction
CL	Cathodluminescence
CTEM	Conventional Transmission Electron Microscopy
EDAX	Company selling EDX equipment
EDS	Energy Dispersive (X-Ray) Spectroscopy
EDX	Energy Dispersive X-Ray Spectroscopy
EPMA	Electron Probe Microanalysis (also known as Electron Probe)
ESCA	Electron Spectroscopy for Chemical Analysis
FTIR	Fourier Transform Infra-Red (Spectroscopy)
FT Raman	Fourier Transform Raman Spectroscopy (See Raman)
GC-FTIR	Gas Chromatography FTIR
GIXD	Grazing Incidence X-Ray Diffraction (also known as GIXRD)
HRTEM	High Resolution Transmission Electron Microscopy
IR	Infrared (Spectroscopy)
IRAS	Infrared Reflection Absorption Spectroscopy
KE	Kinetic Energy
LEED	Low-Energy Electron Diffraction
LTEM	Lorentz Transmission Electron Microscopy
Magnetic SIMS	SIMS using a Magnetic Sector Mass Spectrometer (also known as Sector SIMS
OES	Optical Emission Spectroscopy
PISIMS	Post Ionization SIMS
PHD	Photoelectron Diffraction
Q-SIMS	SIMS using a Quadruple Mass Spectrometer
RA	Reflection Absorption (Spectroscopy)
Raman	Raman Spectroscopy
RBS	Rutherford Backscattering Spectrometry
RRS	Resonant Raman Scattering
RS	Raman Scattering

SAD	Selected Area Diffraction
SAM	Scanning Auger Microscopy
SE	Secondary Electron
SEM	Scanning Electron Microscopy
SEMPA	Secondary Electron Microscopy with Polarization Analysis
SERS	Surface Enhanced Raman Spectroscopy
SFM	Scanning Force Microscopy
SIMS	Secondary Ion Mass Spectrometry (Static and Dynamic)
SPM	Scanning Probe Microscopy
STEM	Scanning Transmission Electron Microscopy
STM	Scanning Tunneling Microscopy
TEM	Transmission Electron Microscopy
TGA-FTIR	Thermo Gravimetric Analysis FTIR
TOF-SIMS	SIMS using Time-of-Flight Mass Spectrometer
WDS	Wavelength Dispersive (X-Ray) Spectroscopy
WDX	Wavelength Dispersive X-Ray Spectroscopy
XAS	X-Ray Absorption Spectroscopy
XPS	X-Ray Photoelectron Spectroscopy
XPD	X-Ray Photoelectron Diffraction
XRD	X-Ray Diffraction

Contributors

Guillermo Bozzolo
Analex Corporation
Brookpark, OH

Brian J. Briscoe
Department of Chemical Engineering
Imperial College
London, UK

Bharat Bhushan
Mechanical Engineering Department
Ohio State University
Columbus, OH

Stephen V. Didziulis
The Aerospace Corporation
El Segundo, CA

John Ferrante
National Aeronautics and
Space Administration
Cleveland, OH

William A. Glaeser
Batelle Laboratories
Columbus, OH

Michael R. Hilton
The Aerospace Corporation
El Segundo, CA

Kenneth C. Ludema
Mechanical Engineering Department
University of Michigan
Ann Arbor, MI

T.A. Stolarski
Department of Mechanical Engineering
Brunel University
Uxbridge, Middlesex, UK

Koji Kato
Mechanical Engineering Department
Tohoku University
Sendai, Japan

The Role of Adhesion in Wear

Friction

Magnetic Recording Surfaces

Surface Analysis of Bearings

The Role of Adhesion in Wear

Introduction, Adhesive Wear

Surface Analysis of Bearings

Boundary Lubrication

Friction

Abrasive Wear

1

Introduction

WILLIAM A. GLAESER ·

Although the subject of this series is surface characterization of materials, the field of tribology tends to use surface analysis to determine surface contact mechanisms. This volume demonstrates the surface science involved in tribology and provides a number of examples of the application of surface analysis.

Tribology is a discipline involving the physics, chemistry, and engineering of moving contacting surfaces. Friction, wear, contact fatigue, lubrication, and adhesion are all elements of the field of tribology. Until the development of the Reynolds equation, which presented a mathematical model of hydrodynamic lubrication, tribology was an empirical science. Over a period of about 50 years, an elegant fluid-dynamic model was developed, enabling engineers to design lubricated bearings with very low friction and practically unlimited life. This fluid dynamic system eventually included even gears, ball bearings, and roller bearings (elastohydrodynamics). Hydrodynamic lubrication involves separation of moving contacting surfaces by a pressurized liquid or gaseous film. This condition has little to do with surface science. It falls within the discipline of fluid dynamics.

Bearings and other sliding surfaces in machinery, however, often do not operate under full film lubrication. Some kind of solid contact occurs even under lubrication. K.L. Johnson and others[1-3] advanced the mathematical analysis of contact stress states and of the concept of *asperity contact*. The understanding of surface deformation and of the resulting microstructural changes during sliding and rolling contact has yielded many ideas on the mechanisms of wear.

Prior to this work in modeling real contacts, Bowden and Tabor[4] introduced the concept of asperity contact. Real surfaces, being bumpy on a microscopic scale, make contact at only a few high points, or asperities. This means that the load is supported on a very small total area during contact. The resulting local contact stress is very high, usually resulting in plastic deformation when ductile materials are involved. The harder asperities then penetrate the surface of the softer material. Sliding contact results in plowing or scoring of the softer surface. Surface deformation induces extremely high plastic strain in a thin surface layer. Hydrocarbon films are penetrated and oxides are broken up, allowing close proximity of metal atoms in both surfaces, promoting adhesion. If asperity contacts can bring atom layers to within 0.5 nm of each other, strong bonding takes place, producing asperity cold welding. The interface between the contacting layers of atoms resembles a grain boundary.

Metal surfaces, whether they are lubricated or dry, are covered with native oxides and hydrocarbons which tend to prevent adhesion. Under sliding and rolling contact (light load conditions) surface damage is minimal. Increased loads can penetrate hydrocarbon films and break through oxide layers. Adhesion and surface damage result. However, if a chemisorbed oxygen layer can immediately take the place of the ruptured oxide, adhesion can be inhibited. The loads at which adhesive damage initiates vary considerably depending on the material surface chemistry, near-surface mechanical properties, and the environment. For instance, Welsh[5] found that steel resisted wear damage depending on the type of oxide developed. He conducted unlubricated wear experiments with mild steel and cast iron using a wide range of load levels. As the load was increased, a transition point was reached in which friction increased by orders of magnitude and surface damage was severe. However, when the load was increased more, another transition point occurred in which friction dropped and surface damage ceased. Welsh concluded that frictional heating caused a change in oxide chemistry, producing a tougher oxide at the second transition.

At the time of these experiments, surface analytical tools were not available. Welsh could not explore his thesis and analyze the surface chemistry of his wear specimens. Current state-of-the-art surface analytical equipment could provide a complete description of the surface and near-surface chemistry and probably would reveal other factors not anticipated by Welsh.

Another surface-related phenomenon which occurs in wear is metal or polymer transfer. During adhesive wear, asperity contacts cold weld and shear off because the bond strength at adhesive contacts is greater than the cohesive strength of either of the two mating materials. The softer of the pair will tend to shear off and transfer to the harder surface. Subsequent passes over the transfer region smears out the original lump and produces a thin layer. This layer can contain material from both surfaces. The structure of the layer in metals is very similar to mechanically alloyed materials.[6] In polymeric materials a structure associated with turbulent mixing develops. The discovery of the nature of transfer layers is a recent development. Advanced techniques[7] for analyzing near-surface wear morphology by TEM in metals has contributed much to this advance.

Lubrication improves the resistance of sliding contacts to adhesion and surface damage. Lubricants are formulated to provide several kinds of protection. Under operating conditions where an oil film is penetrated and asperity welding develops, lubricant additives can produce an adsorbed layer of a gel-like material or a soap which helps to support the load. Under severe conditions, additives are used which

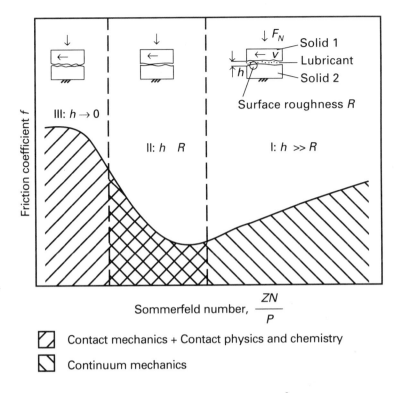

Figure 1.1 Streibeck lubrication curve (from Czichos)[3]

react with the metal surface and produce chemisorbed soft layers that protect the surface.

This process is called *boundary lubrication*. A large percentage of machinery components run under boundary lubrication conditions. The use of surface analytical equipment for the study of the complex processes involved in boundary lubrication is increasing. The hydrodynamic lubrication theory and the boundary lubrication theory can be illustrated by the Streibeck curve shown in Figure 1.1. In the Streibeck curve, friction is plotted against the Sommerfeld number (ZN/P). The quantity Z is the lubricant viscosity, N is the rpm, and P is the bearing pressure. Figure 1.1 is divided into three regions: full-film lubrication (the surfaces are separated by a pressurized film of lubricant), mixed-film lubrication (the load-support film is so thin that some asperity contact occurs, along with wear), and boundary lubrication (friction is no longer influenced by viscosity and the load is supported by a semisolid film). The full-film operating region can be pushed to the left in the diagram by increasing the smoothness of the bearing surfaces, thus allowing thinner films without asperity penetration. Surface roughness analysis (using a profilometer) is an important inspection tool for predicting the reliability of a hydrodynamic bearing system.

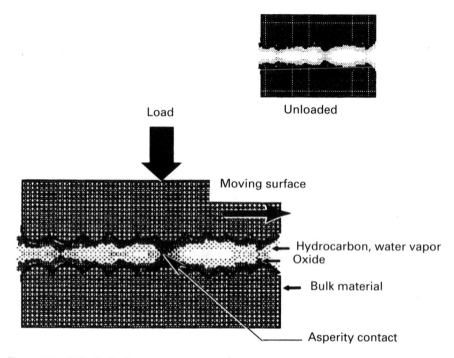

Figure 1.2   Tribological system

Effective boundary lubrication requires as much knowledge of the surface chemistry of the system as possible. This allows the selection of additives to a lubricant and prediction of operating limits of the bearings. This area is not as well advanced because of the complexity of the surface chemistry. Surface analysis contributes more to the development of boundary lubrication models than to inspection or quality control.

It can be seen that tribology deals heavily with surface behavior. Thus, the use of surface analysis is very important to the advancement of the understanding of tribology basics. Except for surface roughness measurements and surface hardness determinations, not much surface characterization of materials for tribology is done on a routine basis. Surface analysis is being adapted to the solution of long-standing problems in the failure of rubbing or rolling surfaces. There is a growing acceptance of surface characterization of materials in this field—led by the magnetic recording industry, where the surface chemistry of magnetic recording media is well defined and inspected by such systems as Auger spectrometers.

The complexities of surface contact are illustrated in Figure 1.2. Two mating surfaces are shown supporting a load. One surface is moving relative to the other. Each surface has a roughness character, so that real contact occurs between asperities. The surfaces also have several layers on them, beginning with a native oxide and proceeding outward to adsorbed species including gases, water vapor, hydrocarbons, and products of chemical reaction. If the system is adequately lubricated, the adsorbed boundary film supports most of the load, allowing few asperity contacts. The two contacting surfaces are surrounded by an environment of liquid lubricants containing chemical additives which react with the surfaces. The lubricants also carry dissolved gases, including oxygen, which may participate in the boundary lubrication reactions. Solid particles (wear debris, dust, microbes, etc.) are also found suspended in the lubricant. The solids may be incorporated in a semisolid boundary film.

In order for the system in Figure 1.2 to work properly, a minimum of solid-to-solid asperity contacts must exist, and the boundary film on the surface must be well bonded, resist penetration, and have low shear strength. The film will be removed by attrition and therefore must be renewable. Additives in the lubricants will renew the film if their reaction rate is faster than the film removal rate. Wear debris must not foul other components in the system (small clearances, capillary tubes, seals, etc.). Some wear debris will coagulate into a "mud" which can be quite disruptive in some systems.

Most boundary additives in modern lubricants are mildly corrosive to the surfaces that they protect from wear and adhesion. Changes in the operating conditions, such as an increase in temperature, can accelerate wear to an unacceptable level.

Additive components such as sulfur can influence oxide formation. It is possible that modification of oxidation during lubrication can contribute to the boundary lubrication process. Sulfur can attack and selectively remove constituents (lead, for instance) in bearing surfaces.

Rubbing contact produces very high strain in near-surface regions. The layer affected is thin—on the order of a few microns. Within this high strain region dislocation density is very high. The dislocation configurations are similar to those found in metal-forming operations. Dislocation cells and subgrains predominate. A typical wear-induced structure in copper is shown in Figure 1.3.

The very high strain in the surface can result in surface chemical changes. Reaction rates are increased and segregation of alloy constituents can occur. In copper–aluminum alloys, lubricated wear can produce selective removal of aluminum from the surface. The result is a thin layer of high copper concentration.[9]

In machining operations, alloy constituents can be lost from a tool by diffusion into the chip rubbing on the tool. Diffusion barrier coatings are used to reduce this effect and increase tool life.

One of the oldest known surface phenomena in tribology is the mechanism of graphite lubrication. During World War II, aircraft flew at higher altitudes than they had flown before. At a critical altitude, electric motor brushes failed. A condition known as *dusting* occurred in which the carbon-graphite brushes would wear at high rates and disappear in clouds of carbon dust. Research into this problem revealed that graphite requires an adsorbed layer of water vapor to lubricate effectively.

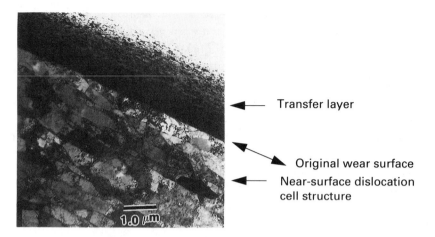

Figure 1.3   TEM micrograph of thin foil section through a wear scar on copper (from Rigney)

When aircraft reached altitudes where the dew point dropped sufficiently that water vapor was not replenished on the brush surfaces, heavy wear of the carbon-graphite resulted. High-altitude brushes were developed in which a constituent was added to take the place of the water vapor. This research was carried out without the benefit of modern surface analytical equipment. It was the result of critical experiments in simulated atmospheres and intuitive analysis of the results.

The use of surface analytical devices in tribological research has been effective in revealing new processes related to wear and friction.[10] Much of the surface chemistry in boundary lubrication is not understood. Therefore, the selection of lubricants has involved empirical methods which are time consuming and expensive. Ideally one would like to be able to select specific chemical systems for a given set of operating conditions on the basis of a set of basic principles.

In addition, the mechanisms of adhesion are being explored, with much ground still to be covered. The effect of the space environment on sliding contact systems has been significant in the use of satellite telemetry, gyro stabilizers, and rocket engine components. Even now, it is suspected that atomic oxygen in space can attack solid lubricant films and possibly disable sensitive instruments.

Electronic recording has produced tribological problems involving surface physics and chemistry. The prevention of head crashes on hard disks involves the use of vapor-deposited protective films that provide lubrication while not inhibiting high density recording. Extensive use of surface analytical equipment in this field has been the norm. Because of the extremely light loads involved in operating read–write heads for magnetic recording disks, new friction-measuring devices are being developed. The concern is to eliminate the surface deformation factor from the friction process and to measure only the shear properties of nanometer films on polymeric surfaces. The atomic force microscope is being adapted for this purpose.

INTRODUCTION   Chapter 1

It is not surprising that surface analytical techniques have been rapidly incorporated into tribological research because of the heavy surface science implications inherent in lubrication, friction, and wear. Most of the applications involving surface science analytical tools have been found in research programs. So far, it has been difficult to convince industry that the investment in such sophisticated equipment is cost effective. However, there are increasing examples of the use of surface analytical equipment.

The importance of surface properties in magnetic recording media has been mentioned above. The magnetic recording industry has been using Auger and XPS routinely for many years. A description of some of the surface analytical methods used in this industry can be found in Chapter 7. A listing of surface science equipment used in tribology and their tribological applications is presented in Table 1.1.

Note that most of the surface analytical equipment listed in Table 1.1 involves high vacuum chambers. Auger and XPS are the techniques most often used in tribological research. Most applications have been in the analysis of dry wear surfaces. These experiments have shown the value of Auger and XPS for uncovering changes in surfaces brought about by sliding contact. Because of the high vacuum, however, organic species, water vapor, and adsorbed gas layers are removed before analysis. Some evidence of residual hydrocarbons has been detected on worn metal surfaces as carbon peaks in the spectra.

In Battelle experiments with Auger *in situ* wear, surface hydrocarbons were detected and were observed to inhibit transfer until they were removed by wear or by sputtering.[11] The surface chemistry of wear debris has been analyzed by XPS measurements. An example of this analysis for wear debris developed in a ball mill setup is shown in Figure 1.4.[12] The figure shows the chemisorbed products resulting from reaction with bulk iron coupons and the reactions that occur when ball milling iron particles are in the presence of the lubricant. The first spectrum is for the ZDDP alone. Comparing peak ratios for phosphorous and sulfur, it can be seen that a phosphorous layer predominates on the solid iron coupon, see spectrum (b). Sulfur and iron predominate on the ball milled wear debris spectrum (c), and energy shifts indicate that sulfide is present on the debris surface.

Ellipsometery and Raman operate in air environment. Ellipsometry has been used to measure the thickness and structure of hydrocarbon boundary films developed during boundary lubrication.[13] An example is described in Chapter 6. Ellipsometery has also been used in the investigation of the role of metal oxides in wear control.[14]

This book has been organized to cover various aspects of tribology that are related to surface science. Chapters by experts include the theory of adhesion, friction between moving surfaces in contact, adhesive wear, abrasive wear, boundary lubrication or lubrication by thin solid semisolid films, and the singular relation of surface science to the performance of mechanical magnetic media data storage systems (hard disks, for instance).

Analytical method	Maximum depth of penetration	Spatial resolution	Tribological applications
Auger	0.5–1.5 nm	1000 nm	Metal transfer, surface segregation Magnetic recording surface inspection Analysis in vacuum
XPS	1.5–7.5 nm	1–4 mm	Lubricant reaction products chemistry Analysis in vacuum
Ellipsometry	400–500 nm	1 mm	Transparent solid film thickness Analysis in air
Raman	400–500 nm	10 nm	Organic film thickness and chemistry Analysis in air
Rutherford backscattering	2 µm	10 nm	Surface film thickness composition
SIMS	Sputters 0.5 nm	500 nm	Surface chemistry Analysis in vacuum
In situ SEM wear test		500 nm	Microstructure of developing wear scar Elemental analysis of transfer products as they are produced Analysis in vacuum
Light microscopy		200 nm	Character and size of wear topography surface film color Nomarski texture
Fourier transform infrared spectroscopy		2000 nm	Chemistry of organic films: polymer transfer and boundary lubrication Analysis in air
Atomic force microscopy		0.1 nm	Friction forces on the atomic level Atomic level surface roughness
Profilometer		0.1–25 µm	Surface roughness by stylus Micro-topography of surface finish and wear

Table 1.1   Surface analytical tools used in tribology

Chapters 2 and 3 describe the theory of adhesion and friction, showing the origins of forces between contacting asperities at the atomic level. Both of these chapters lead one to believe that measurements of friction and adhesion on the atomic level should be useful in testing the theories.

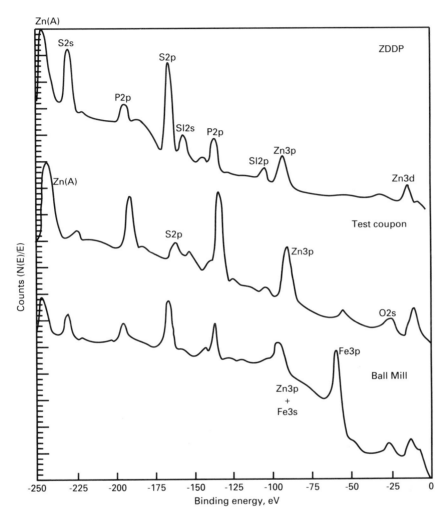

Figure 1.4 Low binding energy XPS spectra for three conditions of iron reaction with the boundary lubricant zinc dialkyldithiophosphate (ZDDP)

A number of investigators have been using atomic force microscopy to measure friction forces on an atomic level and to study the relation between atomic binding energies and interatomic separations.

Mate[15] and colleagues used an atomic force microscope in their studies of friction, sliding the tungsten wire tip over the basal plane of a single graphite crystal in polycrystalline graphite at low loads. They found a friction force influenced by the atomic structure of the graphite surface.

Kaneko[16] investigated friction variations on magnetic storage disk surfaces. He used a modified atomic force microscope with a leaf spring friction force measuring system. He found friction variations with periods of 50–60 nm.

Ferrante and colleagues[17] used the atomic force microscope to determine the force–distance characteristics of approaching surfaces of pure metals to verify these characteristics as derived from theory. They found that the force–distance curve is influenced by the geometry of the probe tip. However, the shape of the force–distance curve is universal.

Landman and colleagues[18] used the atomic force microscope to verify computer-generated models of asperity contact involving approximately 50 atoms.

This sampling of the current usage of the atomic force microscope in tribology to supplement the development of theoretical models for adhesion and friction shows the probable future direction of fundamental research in this area. This research is barely in its infancy. The relation between friction on an atomic level and macroscopic friction processes is yet to be developed and will require considerable investigation before any breakthrough comes.

Kaneko's work suggests possible future applications of ATM for characterization of magnetic recording material.

Briscoe[19] has recognized that the friction developed in polymer films on rigid surfaces is a function of the shear properties of the solid films. The shear properties of the polymer films are influenced by the interaction of polymer molecules. These interactions are more complex than metal shear processes. The resistance to flow arises from continual deformation of the polymer molecules, which on relaxation convert the stress into thermal energy. The details of this process are influenced by the molecular structure and contact pressure in a film. Characterization of films using vibrational spectroscopy is being investigated as a tool to improve understanding of the basic mechanisms of friction for polymer surface films. This might be extended to transfer films (PTFE and UHMWPE) on metal surfaces.

Chapter 4 describes the application of adhesion theory to adhesive wear phenomena. Surface analytical techniques are used to detect small amounts of metal transfer during sliding contact. Of critical importance to understanding the adhesive wear process and to developing methods for selection of compatible materials for sliding contact, is the determination of the surface conditions favorable to metal transfer. Does the native oxide have to be removed before transfer can occur? Some oxides appear to promote the transfer process.

Kato, in Chapter 5, describes the application of *in situ* SEM wear experiments to the development of a theory of abrasive wear. *In situ* wear experiments in the SEM show the microscopic deformation processes that occur during wear. The change in surface morphology can be followed in a selected small area for consecutive passes of a sliding contact. The abrasion process has been followed in real time and recorded on video tape.[20]

In Chapter 6, Ludema demonstrates the application of ellipsometry to the development of a theory for boundary lubrication.

Bushan, in Chapter 7, and Hilton, in Chapter 8, describe the use of surface analytical equipment for routine inspection of mechanical components—Bushan for magnetic recording media, and Hilton for precision rolling contact bearing surfaces.

Surface analysis has been an important part of tribological research. To date, the greater use of surface analytical equipment for tribological purposes (save for profilometry and scanning electron microscopy) has been in the research laboratory. That is, routine surface characterization of bearings, gears, seals, and other manufactured sliding or rolling surfaces is not currently prevalent. However, in certain industries—especially in the magnetic recording industry—the necessity for maintaining defect-free recording surfaces has required the use of sophisticated surface analysis as routine quality control. In addition, the development of new surface treatments like ion implantation and CVD diamond coatings requires surface characterization in order to assess the integrity of the treatments.

Expensive surface analytical equipment is found mostly in research laboratories and in specialized industries like the semiconductor, magnetic recording, and advanced surface treatment industries. In the future, it is expected that surface characterization will become more widely used for quality control and for failure analysis. Companies will find that investment in surface analytical equipment will become more cost effective, starting with companies involved in the manufacture of precision components.

It is expected that surface characterization will be used more in the selection of materials and lubricants for mechanisms of improved reliability and life. An example of the potential for solving an engine lubrication problem is found in research by Mattsson et al.[21] These researchers found that when alcohol was substituted for conventional diesel fuel, it reduced the beneficial boundary films developed on the cylinder liner surfaces and increased the wear of piston rings and cylinders. Wear tests and engine tests revealed the adverse effect of alcohol fuel on lubricant additives in engine performance. The actual cause of the increased wear was found by the use of post test analysis of cylinder surfaces by XPS analysis.

Development of new materials for tribology will involve an increased emphasis on surface treatment rather than new alloys or monolithic ceramics. In the field of polymers, the incorporation of lubricants and strengthening fibers in composites is increasing. For both of these material development trends, surface characterization will be essential. The atomic force microscope and the scanning tunneling microscope developed by physicists and materials scientists will be adapted to characterize real materials on the atomic scale. Just as light microscopy was adapted to the study of the crystalline structure of materials, ATF and STM microscopy will find their way into the metallographic laboratories. The imaging of surface atom configurations will be useful in designing coatings with required bonding strengths. These microscopes can be used at "low magnification" where micron-sized crystals on a surface can be resolved in three dimensions. The surface structure of diamond and diamond-like coatings deposited under different conditions are strikingly apparent in AFM micrographs in a 3000-nm square area.[22] These micrographs can be used to understand Raman spectra taken from the same surface. Comparing a polished and etched sample of AISI 1018 steel using light microscopy and AFM has shown that, contrary to accepted beliefs, grain boundaries etch faster than grains—the ferrite in the pearlite is the slower etching constituent. It has been speculated that carbon atoms migrate to grain boundaries, forming cementite between the ferrite grains. These particles serve as barriers to dislocation movement. This might explain the Hall–Petch effect, in which the smaller the grain size, the stronger the steel.

This example shows an exciting new approach to the study of material structure and its relation to mechanical behavior and to wear. It is likely that AFM and STM will become powerful adjuncts to the techniques for understanding of material mechanical behavior. In addition, these and other surface analytical devices will be used to help tailor surface treatments and lubricants for tribological applications. Just as the SEM started out as an interesting laboratory innovation and ended up an essential component in the characterization of surface finish and wear topography, so should ATM and STM soon be accepted in the metallographic laboratories.

References

1 K.L. Johnson, *Contact Mechanics*, Cambridge University Press, 1985.

2 F.P. Bowden and D. Tabor, *The Friction and Lubrication of Solids*, Oxford Clarendon Press, 1964, 29–51.

3 J.A. Greenwood and J.B.P. Williamson, Contact of nominally flat surfaces, *Proc. Roy. Soc.*, **A295**, 300–319, 1966.

4 Ibid Bowden and Tabor.

5 N.C. Welsh, The dry wear of steels, *Phil. Trans. Roy. Soc.*, **A257**, 31–70, 1965.

6 H.J. Fecht, Z. Han and W. Johnson, Metastable phase formation, in Zr–Al system by mechanical alloying, *J. Applied Physics*, **67** (4), 1744–48, 1990.

7 P. Heilmann, J. Don, T.C. Sun, W.A. Glaeser, and D.A. Rigney, Sliding wear and transfer, Proceedings *Wear of Materials 1983*, ASME, 414–425.

8 Czichos, H., *Tribology*, Elsevier Tribology Series 1, 1978, 131.

9 M.T. Thomas and W.A. Glaeser, Surface chemistry of wear scars, *J. Vac. Sci. Technol.*, **A2**(2) 1097–1101, 1984.

10 W.A. Glaeser, The use of surface analysis techniques in the study of wear, *Wear*, **100**, 477–487, 1984.

11 W.A. Glaeser, Transfer of lead from leaded bronze during sliding contact, *Proceedings of Wear of Materials*, 1989, 255–260.

12 W.A. Glaeser, Ball mill simulation of wear debris attrition, *Proceedings of 18th Leeds-Lyon Symposium*, 1991.

13 C.M. Allen, E. Drauglis, and W.A. Glaeser, Aircraft propulsion lubricating films additives: Boundary lubricant surface films, *AFAPL-TR-73-121*, **3**, June 1976.

14 S.C. Kang and K.C. Ludema, The role of oxides in the prevention of scuffing, *Proceedings Leeds-Lyon Symposium*, Butterworths, 1984, 3–7.

15 C.M. Mate, C. Mathew, G.M. McClelland, R. Erlandsson, and S. Chang, Atom-scale friction of a tungsten tip on a graphite surface, *Phys. Rev. Lets.*, **59** (17), 1942–1945, 1987.

16 R. Kaneka, A frictional force microscope controlled with an electromagnet, *J. of Microscopy*, **152**, 363–369, 1988.

17 A. Banerjea, J.R. Smith, and J. Ferrante, Universal aspects of brittle fracture, adhesion and atomic force microscopy, *Proc. Mater. Soc. Symp.*, **140**, 89–100, 1989.

18 U. Landman, W.D. Luedtke, N. Burnham, and R.J. Colton, Atomistic mechanisms and dynamics of adhesion, indentation and fracture, *Science*, **248**, 454–461, 1990.

19 B.J. Briscoe, P.S. Thomas, and D.R. Williams, Microscopic origins of the interface friction of organic films: The potential of vibrational spectroscopy, *Wear*, **153**, 263–275, 1992.

20 K. Kato, D.F. Diao, and M. Tsutsumi, The wear mechanism of ceramic coating film in repeated sliding friction, *Proceedings of Wear of Materials*, ASME, 243–248, 1991.

21 L. Mattsson, H. Abramsson, and B. Olsson, Surface spectroscopic study of reaction layers in alcohol-fuelled diesel engines, *Wear*, **130**, 137–150, 1989.

22 T.L. Altshuder, Atomic-scale materials characterization, *Advanced Materials, and Processes*, Sept., 18–21, 1991.

2

The Role of Adhesion in Wear

JOHN FERRANTE and GUILLERMO BOZZOLO

Contents

2.1 Introduction

Tribology is a field that spans a wide range of interests, from the engineer who has to make a particular piece of equipment work, to the scientist who is interested in the underlying mechanisms which produce wear and friction. It is quite complicated in that one is dealing with phenomena that involve widely different materials and environments, often in situations where characterization of experimental conditions is difficult. This chapter is concerned with the more fundamental aspects of tribology. From the scientist's viewpoint, the dominating processes involve attractive forces that exist between the atomic species present, the sum of which dominates all of the macroscopic phenomena observed.

An underlying process in many phenomena involved in wear or friction is adhesion. We will define adhesion as the attractive force across an interface. An interface is defined as two surfaces in contact. These interactions give rise to all of the phenomena in tribology. Wear ultimately involves a separation at some interface—either internally or between the interacting surfaces such as at a bearing. Friction also involves the forces across some interface between moving components. Lubrication consists of finding materials which reduce these forces.[1-6] This chapter will be concerned with the experimental and theoretical techniques used to understand the adhesive interaction. We will consider mainly static rather than dynamic interactions such as the energy dissipation mechanism involved in friction. We will first outline the techniques involved in measuring the adhesive force and the difficulties involved therein, then present experimental evidence for the atomic nature of such forces, and finally, present theoretical approaches to adhesion from an atomic standpoint.

2.2 Considerations for Experiments

Background

Adhesion is evident in tribology when transfer of material occurs from one surface to another, as Rigney[6] points out in a particularly clear discussion of the topic. There is an interest in measuring adhesive forces for a wide range of applications besides tribology in order to design stronger or (in the case of ceramic, fiber-reinforced composites) weaker interfaces.

Consequently, a number of experimental techniques have been designed to measure adhesion, see Table 2.1. Upon examining these approaches it is apparent that there is considerable ambiguity in obtaining information concerning the fundamental forces of interaction at an interface. In most practical applications the objective is to obtain a bond strong enough to sustain the conditions experienced during use. However, this does not guarantee that the bond or material will not fail under normal operation since at best what is obtained is statistical information. One objective of more fundamental studies is to set bounds based on physical principles which will permit the user to know the maximum attainable strength.

We now concentrate on examining a simple experiment to measure adhesion with implications in tribology in order to elucidate difficulties in interpretation. Consider an idealized experiment, a single asperity in contact with a flat surface— an experiment which may now be performed with the atomic force microscope (AFM).[7] The object of the discussion is to determine what information can be obtained about the interfacial force from a loading and pull-off experiment. In Figure 2.1a we show a standard stress–strain curve. We can see from Figure 2.1b that the results of the experiment depend strongly on the properties of the materials and the adhesive force. If the solids are elastic without interfacial adhesion, then a simple Hooke's law behavior would be observed upon loading and unloading, whereas with adhesion the shape of the curve would be the same but a negative load would have to be applied to obtain separation in order to overcome the adhesive force. If the interface is stronger than one or both of the materials, plastic deformation and ductile extension may occur in one or both materials, and fracture will occur in the weaker of the two materials. Thus, in either case, the experimental results are a combination of the mechanical behavior of the bulks, and if the interface is stronger no information is obtained about the strength of the interface.

Method	Principle
Bending	Substrate bent or twisted until film removed
Squashing	Substrate squashed until film removed
Abrasion	Burnishing or abrasion of surface to remove film
Heating and quenching	Heating and sudden quenching will cause film to be removed because of stresses developed by thermal expansion and contraction
Scratching	Film scratched through by a probe. Alternatively, parallel grooves cut into film with decreasing separation until intervening material lifts from substrate
Hammering	Hammering breaks up and removes film
Indentation	Substrate indented from side opposite to film. Coating examined for cracking of flaking off at various stages of indent formation
Pulling	Film pulled off directly if it is thick enough. If not, backing attached using solder, adhesives, or electroforming
Peeling	Film peeled off using a backing of adhesive tape or electroplated coating
Deceleration	The film and substrate are subject to violent deceleration, which removes the film. Various experimental arrangements are possible: coated bullet stopped by steel plate, ultracentrifuge, and ultrasonic vibration
Electromagnetic	Applying a force to a conducting film by a $v \times B$ force
Shock wave	Pulsing the backface of a specimen with a laser
Blistering	Film is deposited so that no adhesion exists over a particular area. Air is then introduced into this area and the pressure at which film starts to lift from the area of no adhesion is measured

Table 2.1   Engineering techniques for measuring adhesion[28]

Adding the complexities of the mechanical properties of the testing machine and time and temperature dependence, and the fact that the stress distribution at the interface may not be uniform, it is clear that the interpretation of even the simplest of the experiments is not straightforward in terms of the adhesive strength of an interface. Furthermore, even the locus of failure may not be obvious, and techniques such as scanning electron microscopy or Auger electron spectroscopy (AES)[3, 5] may be needed to identify the locus of failure. Each of the tests presented in Table 2.1 suffer from similar ambiguities in interpretation with regard to determining interfacial adhesive strength.

THE ROLE OF ADHESION IN WEAR   Chapter 2

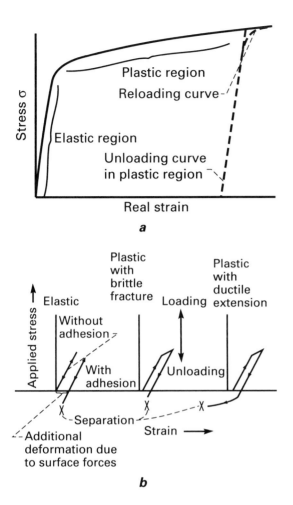

Figure 2.1 Stress–strain diagram for a tensile test: a) typical stress–strain diagram, b) loading and unloading stress–strain curves with adhesion (idealized single contact)

Macroscopic Experiments

We now present the results of experiments performed under very well controlled conditions to emphasize the complexities of interpretation. Pepper[8] placed a pin-on-disk system in an ultrahigh vacuum (UHV) system. An Auger electron spectrometer which has fraction-of-monolayer sensitivity[3, 5] was used to detect transfer. The surfaces were atomically cleaned by sputtering, prior to performing transfer experiments where Fe, Ni, or Co pins were rubbed against W, Ta, Mo, or Nb disks. The model that the cohesively weaker metal transfers to the cohesively stronger metal as discussed by Chen and Rigney[9] would predict that the softer metal would

Disk	Rider	Transfer of metal from rider to disk
Tungsten	Iron	Yes
	Nickel	Yes
	Cobalt	Yes
Tantalum	Iron	No
	Nickel	No
	Cobalt	Yes
Molybdenum	Iron	No
	Nickel	No
	Cobalt	Yes
Niobium	Iron	No
	Nickel	No
	Cobalt	Yes

Table 2.2   Metallic transfer for dissimilar metals in sliding contact

transfer to the refractory metal. Table 2.2 shows that this indeed occurred for the tungsten disk, but for the other disks only cobalt transferred in all cases. Pepper concluded that the transfer was dominated by a change in the mechanical properties of the softer metals in that Fe and Ni strain-hardened under the high load whereas Co, which has a different crystal structure (hexagonal close packed) and easy slip systems, transferred in all cases to the harder material. Chen and Rigney point out that transfer should occur from disc to pin except when the disc is much harder than the pin, which was the case in Pepper's experiments. These experiments emphasize that the results of wear experiments cannot be predicted on the basis of simple arguments and that other factors must be taken into account.

Atomic Level Experiments

This section addresses the effects of atomic-level adsorption on the adhesive force. *Lubricants* probe interfacial forces by altering them. These atomic lubricating effects are differentiated from *macroscopic* lubricants (solid and liquid) which may have little to do with adhesion at the solid–solid interface; e.g., for solid lubricants, shear may occur in the intervening film, and with liquids hydrodynamic effects can occur.[10] The effects of adsorption on metal–metal adhesion are discussed first. Wheeler[11] used a pin-on-flat configuration in order to examine the effects of the adsorption of oxygen and chlorine on the interfacial shear strength (coefficient of static friction) of iron, steel, and copper contacts at partial atomic layer concentrations. The experiments were performed in UHV with AES to examine surface coverage. Figure 2.2 shows the results of the study. It is evident that partial monolayer

THE ROLE OF ADHESION IN WEAR   Chapter 2

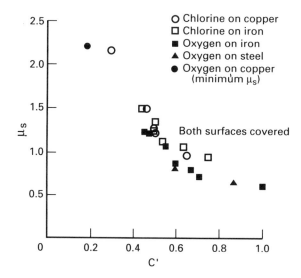

Figure 2.2 Effects of oxygen and chlorine adsorption on static friction of metal–metal contact: Static coefficient of friction μ_s as a function of adsorbate concentration C'

concentrations of oxygen or chlorine reduce the interfacial shear strength of the interface. These results can be interpreted in terms of the adhesion at the interface. Bowden and Tabor[1, 2] in their classic work point out that when clean metal surfaces are in contact and a shear force is applied, the combination of load and strong interfacial bonding forces causes plastic deformation, and therefore junction growth occurs (the junction adheres and the asperity flows). Wheeler's experiments imply that adsorption at the interface reduces this interfacial bonding force. These static friction experiments are likely more related to the wear process than to friction since shear at the interface or near is fundamental to transfer, which is a wear phenomenon. Recently, Gellman[12] has performed similar experiments on single crystal nickel surfaces covered with half a monolayer of sulfur on which ethanol was adsorbed at partial monolayer and multilayer coverages. The sulfur at these coverages did not have an apparent lubricant effect and was adsorbed to prevent decomposition of the ethanol by the clean nickel surface. Gellman found results similar to Wheeler's: at low ethanol coverages stick (and thus conditions necessary for transfer) was observed, at fractional monolayer coverages stick–slip was obtained, and at monolayer coverages continuous slip as in a dynamic friction experiment occurred. We note however, that Gellman's experiments were performed on single crystals and Wheeler's were performed on polycrystalline surfaces. It is apparent that with both wear and friction, atomic-level phenomena can control the observed behavior.

Some similar experiments on metals in contact with ceramics present some surprising results. Pepper,[13] using a ball-on-flat configuration, measured the interfacial

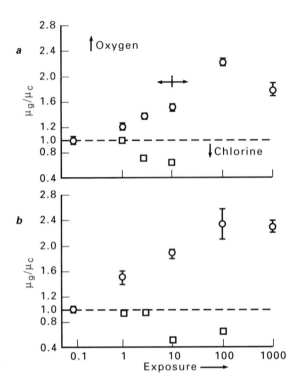

Figure 2.3 **Shear coefficient of oxygenated and chlorinated metals in contact with clean (0001) sapphire: a) nickel, b) copper**

shear strength of Ag, Cu, Ni, and Fe spheres in contact with a single crystal sapphire flat in an ultrahigh vacuum system. First, he found that the shear strength depended on the metal. Surprisingly, he also found that for unlike metal–metal contacts, oxygen adsorption at fractional monolayer coverages increased the interfacial shear strength, whereas chlorine decreased it (Figure 2.3). Ethylene also increased the interfacial shear strength. Again, atomic-level adsorption had macroscopic effects. These results could also be produced in macroscopic pin-on-disk experiments in UHV. Thus adsorbates that are potential wear reducers with metals can be wear promoters with ceramics.

Microscopic Contacts

We note that it is now possible to measure these atomic level forces with the AFM.[7] Recently, Duerig et al.[14] have measured the adhesive force at metal tip-flat interfaces and a function of distance. Mate et al.[7] have used the AFM to measure friction forces at the atomic-level. Therefore, there is a new realm arising in tribology where atomic level phenomena can be probed directly.

THE ROLE OF ADHESION IN WEAR   Chapter 2

Figure 2.4   Four-degree twist boundary for an fcc (001) interface

2.3 Theoretical Considerations at the Atomic Level

Background for Theory

Just as with experiments, theory has reached the point where atomic effects can also be probed. Until recently, the bulk of theoretical research in tribology has involved continuum mechanics approaches. This occurred because the geometric and material complexities in real applications had proven intractable for more fundamental approaches, such as solving the Schroedinger equation. This is in part because solutions to the Schroedinger equation in a solid relied on the periodicity of the lattice to simplify the solution.[15] A real interface is extremely complex. For example we take a very simple situation in which we rotate a single crystal surface of a face-centered metal by 4° at a (100) plane. It is clear from Figure 2.4 that this simple defect in the solid creates an exceedingly complex situation. Furthermore, this rigid geometry is not the real situation, which involves a slight relaxation of all of the atoms to their minimum energy configuration. Thus, a search would have to be done solving the Schroedinger equation for a multitude of configurations in the hope that the minimum energy structure could be found. These are the simplest considerations; many other complexities are involved in real applications, such as motion of the interfaces and temperature.

In spite of these seemingly insurmountable difficulties, the situation has improved in both fundamental techniques and methods developed from them.

First, for quantum mechanical calculations the symmetry problem has been mitigated by a technique referred to as *super cells*[15] in which the structure of a real defect is represented by a large cell that is repeated throughout the solid. The idea is that a big enough cell might converge to the energetics of the real defect. In addition, new techniques have been developed in which time and temperature dependence can be included.[16] These improvements have been made possible largely by increases in computer speed. Unfortunately, these improvements are not sufficient to make the calculations in practical tribology tractable, but they are an important tool for progress in atomic simulations, as will be discussed.

Because of the difficulties mentioned, other techniques have been developed to treat problems in material science and tribology. In the past, the principal approach used has been pair-potentials. These are two-body interactions, for example, the Coulomb interaction. The procedure used is to first specify the geometry of a defect, then sum over all two-body forces in order to calculate the energy of a given defect. The problem is that pair-potentials do not properly take into account the interactions occurring in most materials.[17] For metals, the electron gas has some properties similar to a fluid and there is a relaxation which is not treated properly by pair-potentials.[18] In covalently bonded solids, there are contributions to the energy from bond bending (so called many-body contributions[19]) which are not taken into account by pair-potentials. The principal advantage of pair-potentials has been their simplicity, and in addition, there has been little else to turn to. There have been attempts to construct many-body potentials[19] to treat such problems. Recently, new approaches with a stronger physical basis have been developed. Two of these are discussed: the embedded atom method (EAM)[20] and equivalent crystal theory (ECT).[21]

Universal Binding Energy Relation

Before proceeding with a description of these two semiempirical methods, it is necessary to introduce some results of Rose, Smith, and Ferrante (RSF)[22] concerning the properties of binding energy curves. First, we show the results of a first-principles, metallic adhesion calculation performed by Ferrante and Smith.[23] They used first-principles methods to calculate the binding energy as a function of separation at an interface between two metals (adhesion energy). Some results for a number of different metals in contact are shown in Figure 2.5. An important conclusion from these calculations is that the adhesion energy (the value at the minimum in the curve) can be stronger than the weaker of the two materials, as was postulated earlier with regard to transfer. RSF found that these curves could be scaled such that they would fall onto one curve (Figure 2.6), thus indicating that the shape of the curves is the same for all of these combinations of metals in contact. Stated more formally, the energy versus separation can be written as

$$E(r) = \Delta E \, E^*(a^*) \tag{2.1}$$

THE ROLE OF ADHESION IN WEAR Chapter 2

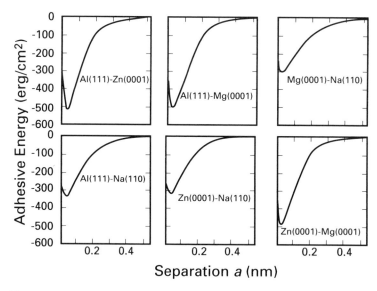

Figure 2.5   Adhesive binding energy curves for the contacts shown

where

$$a^* = ((r - r_e) / l) \tag{2.2}$$

ΔE is the binding energy, a^* is a scaled length, r is the distance between surfaces, r_e is the equilibrium separation and l is a scaling length given by

$$l = (\Delta E / (d^2 E / dr^2) r_e)^{-1/2} \tag{2.3}$$

RSF found that the scaled energy could be represented by a simple functional for first proposed by Rydberg, for diatomic molecules:

$$E^* (a^*) = -(l + a^*)^{-a^*} \tag{2.4}$$

Further, they found that this result was more general than just for adhesion and could also be applied to diatomic molecules, chemisorption, and cohesion. Therefore this result was named the universal binding energy relation (UBER); it will be used in the subsequent discussion.

Semiempirical Methods

EAM is based on first-principles methods that give the energy to embed an ion into jellium (an idealization of the solid where the positive ions are treated as a uniform distribution of positive charge). It has been developed primarily for metals and includes the contributions from redistribution, but not relaxation, of the electron

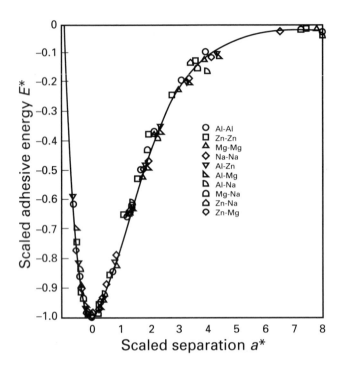

Figure 2.6   Scaled adhesive binding energy curves for the contact shown in Figure 2.5

gas. The procedure involves specifying the geometry of a defect, for example a grain boundary, and obtaining the electron density for a given ion in the defect by overlapping the atomic electron densities of all of the other atoms in the solid at that site. The energy of the atom in the defect is then obtained from a simple expression containing two terms, with one term depending on the electron density at the site due to all of the other particles, and a pair repulsion term for which the parameters are known. The UBER is used to define the expression for the energy. This procedure is carried out for each atom that defines the defect and then the contributions are summed to get the total energy. EAM is not much more difficult to use than pair-potentials, yet it corrects many of the deficiencies of pair-potentials outlined above. It has been successfully applied to a number of problems, e.g., phonons, liquid metals, defects, fracture, surface structure, and surface segregation.

We also present the results of a newer technique, ECT. ECT also relies on the UBER. The energy of an atom in a defect is equal to the energy of an atom in an isotropically expanded or contracted perfect crystal (the equivalent crystal). The energy difference between the atom in the defect and that in the perfect crystal is treated by perturbation theory. The condition used is to find the lattice parameter of the expanded or contracted equivalent crystal that gives a zero value for the perturbation and thus leaves only a simple expression for the UBER to evaluate. The perturbation equation to be solved for the lattice parameter is a simple transcendental equation. The calculation proceeds in the fashion outlined for EAM, i.e., summing the energies over all nonequivalent sites. Again, the procedure is almost as simple as pair-potentials to implement. ECT has certain advantages in that it can be simply applied to bcc as well as fcc metals, and also to covalently bonded solids.

We now show a comparison of EAM and ECT for calculation of surface energies of metals with first-principles results which are the standards for testing in Table 2.3 (surface energies for a clean metal surface are very difficult to obtain experimentally.) We can see that EAM underestimates these values by 40%, whereas ECT agrees to within 10%. The free surface is a particularly difficult test of the methods because it is a large perturbation from the bulk. The deficiency of EAM for this case probably arises from overlapping atomic electron densities which is a better for internal interfaces where the perturbation isn't as large. ECT, to a degree, includes relaxation of the electron density in its procedure. In Figure 2.7 we show[24] some adhesion curves calculated using ECT. These are trivial to obtain compared to first-principles techniques discussed previously. These calculations could be performed on a personal computer for either EAM or ECT.

Element	Crystal face	ECT rigid	ECT relaxed	LDA rigid	Other
Cu	(111)	1830	1780	2100	1170
	(100)	2380	2320	2300	1280
	(110)	2270	2210		1400
Ag	(111)	1270	1230		620
	(100)	1630	1600	1650	705
	(110)	1540	1510		770
Ni	(111)	2400	2320		1450
	(100)	3120	3040	3050	1580
	(110)	2980	2910		1730
Al	(111)	920	860		
	(100)	1290	1220		
	(110)	1310	1230	1100	
Fe	(110)	1820			
	(100)	3490		3100	1693
W	(110)	3330			
	(100)	5880		5200	2926

Table 2.3 **Adhesive binding energies for the systems shown calculated by ECT, LDA (first-principles), and EAM (other)[21]**

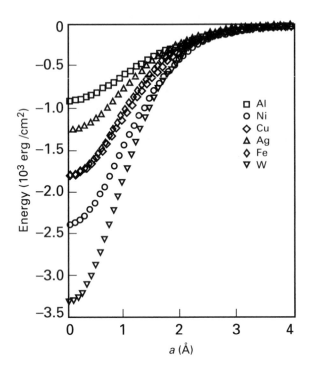

Figure 2.7 Variation of adhesive binding energy with separation for the lowest index interface between Al, Ni, Cu, Ag, Fe, and W

We present another example of a calculation related to wear using ECT, namely slip at a (100) interface[25] in an fcc crystal. This initial calculation was performed with zero applied load, and the two halves of the crystal were held fixed in geometry. These conditions were imposed in order to examine whether certain scaling relations exist for slip, but are not necessary. Figure 2.8 shows the results for sliding in two different directions on a solid surface. The solid line is a fit of the results to a Fourier series. It can be seen that the curves scale. It was discovered that the scaling parameters are the lattice parameter for length (as might be expected) and the surface energy for the energy. Thus some simple underlying properties can be discovered from such calculations. It is possible to include dynamic processes, such as transfer, in defect calculations by use of molecular dynamics techniques. Molecular dynamics involves solving Newton's second law based on a potential obtained from the methods outlined. Landman et al.[26] have performed extensive studies of tip-flat surface interactions showing dynamic adhesion and transfer using EAM for metals that exhibit many properties of interest in wear.[27]

In conclusion, calculational theory has advanced to the point where phenomena in wear can be examined at the atomic level. The procedure of choice will be to develop semiempirical techniques guided by first-principles methods and then to apply these techniques to problems of practical interest.

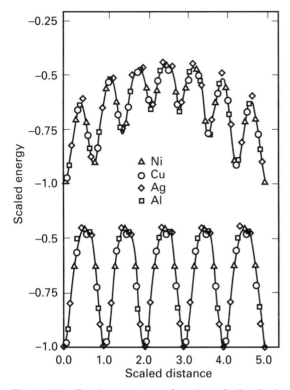

Figure 2.8 Total energy as a function of slip displacement in (100) interfaces of Ni, Cu, Ag, and Al at zero load

2.4 Conclusions

The study of adhesion and wear remains a difficult experimental problem at the macroscopic level. However, studies of these phenomena at the microscopic level in well-controlled environments, and the use of surface analytical tools, along with studies using the AFM show considerable promise. Theory has advanced to the point where atomic simulations of phenomena of interest in adhesion and wear are feasible. It is possible to obtain an understanding of friction and wear using both experimental and theoretical techniques presently available.

References

1 F.P. Bowden and D. Tabor, *The Friction and Lubrication of Solids*, Oxford University Press, Oxford, 1950.

2 F.P. Bowden and D. Tabor, *The Friction and Lubrication of Solids II*, Oxford University Press, Oxford, 1964.

3 D.H. Buckley, *Surface Effects in Adhesion, Friction, Wear and Lubrication*, Elsevier, Amsterdam, 1982.

4 E. Rabinowicz, *Friction and Wear of Materials*, John Wiley and Sons, New York, 1965.

5 T.F.J. Quinn, *Physical Analysis for Tribology*, Cambridge University Press, Cambridge, 1991.

6 D.A. Rigney, *Scripta Met. et Mat.*, **24**, 799, 1990.

7 C.M. Mate, G.M. McClelland, R. Erlandsson, and S. Chiang, *Phys. Rev. Lett.*, **59**, 930, 1987.

8 S.V. Pepper and D.H. Buckley, *NASA TND-6497*, 1971.

9 L.H. Chen and D.A. Rigney, *Wear*, **136**, 223, 1990.

10 J. Ferrante and S.V. Pepper, *Mat. Res. Soc. Symp. Proc.*, **140**, 37, 1989.

11 D.R. Wheeler, *J. Appl. Phys.*, **47**, 1123, 1976.

12 A.J. Gellman, *J. Vac. Sci. Tech.*, **A10**, 180, 1992.

13 S.V. Pepper, *J. Appl. Phys.*, **50**, 8062, 1982; see also *J. Vac. Sci. Tech.*, **20**, 643, 1982.

14 U. Duerig, O. Zueger and D.W. Pohl, *Phys. Rev. Lett.*, **65**, 349, 1990; see also, *J. of Microscopy*, **152**, 259, 1988.

15 N.W. Ashcroft and N.D. Mermin, *Solid State Physics*, Holt, Rinehart and Wilson, New York, 1976.

16 D.P. Vicenzo, D.L. Alehand, M. Schlutter, and J.W. Wilkins, *Phys. Rev. Lett.*, **56**, 1925, 1986; see also, M.W. Payne, P.D. Bristowe, and J.D. Johannapoulis, *Phys. Rev. Lett.*, **58**, 1348, 1987.

17 A. Carlsson and N.W. Ashcroft, *Phys. Rev.*, **B27**, 2101, 1983.

18 J.R. Smith and J. Ferrante, *Phys. Rev.*, **B34**, 2238, 1986.

19 F.H. Stillinger and T.A. Weber, *Phys. Rev.*, **B31**, 5262, 1982.

20 S.M. Foiles, M.I. Baskes, and M.S. Daw, *Phys Rev*, **B33**, 7983, 1986.

21 J.R. Smith, T.A. Perry, A. Banerjea, J. Ferrante, and G. Bozzolo, *Phys. Rev.*, **B44**, 6444, 1991.

22 J.H. Rose, J.R. Smith, and J. Ferrante, *Phys. Rev.*, **B28**, 1935 (1983); see also, *Phys. Rev. Lett.*, **47**, 675, 1983.

23 J. Ferrante and J.R. Smith, *Phys. Rev.*, **B31**, 3427, 1985.

24 A. Banerjea, J. Ferrante, and J.R. Smith, *Fundamentals of Adhesion*, (L.H. Lee, Ed.) Plenum, New York, 1990, p. 325; see also, *J. Phys. Cond. Mat.*, **2**, 8841, 1990.

25 G. Bozzolo, J. Ferrante, and J. R. Smith, *Scripta Met. et Mat.*, **25**, 1927, 1991.

26 U. Landman, W.D. Luedtke, N.A. Burnham, and R.J. Colton, *Science*, **24**, 454, 1990; U. Landman and W.D. Luedtke, *J. Vac. Sci. Thechnol.*, **B9**, 414, 1991.

27 J.A. Harrison, C.T. White, R.J. Colton, and D.W. Brenner, *Surf. Sci.*, **271**, 57, 1992.

28 D.S. Campbell, *Handbook of Thin Film Technology,* (L.I. Maissel and R. Glang, Eds.) McGraw Hill Book Co., New York, 1970, Chapter 12.

3

Friction

B.J. BRISCOE and T.A. STOLARSKI

Contents

3.1 Introduction

This chapter outlines certain basic elements of friction mainly in the context of metallic systems. Brief reference is made to polymeric and ceramic systems. It adopts the two-component noninteracting model which, with all its limitations, serves as a useful first-order vehicle to rationalize a most complex subject. The two components—that is, the adhesion component and the ploughing component—are separately described. Two facets, scuffing and intermittent motion, are also reviewed. Brief mention is also made of recent developments in microscopic models, and rather special phenomenon such as Coulombic and Schallamach friction processes are briefly introduced.

In the final conclusions, we reflect upon the current status and further developments in the study and interpretation of friction processes.

3.2 Sliding Friction

Basic Concepts

External friction is a term usually given to all phenomena taking place within the contact zone created by two solids under the action of normal and tangential forces. In general, both the surface and the subsurface regions of contacting bodies are affected by the frictional interactions, and the environment also plays an active role in the processes which occur.

During friction processes, two bodies are generally in contact only over very small discrete areas which generally constitute a tiny fraction of a nominal or geometrical contact area. Real contact over the whole nominal area is only possible for ideally smooth and flat surfaces or with very soft, highly deformable solids; cleaved mica and soft elastomers are two of the classical exceptions. Because surfaces of engineering components are invariable rough, the nominal area of contact between them is not sufficient to define the contact characteristics. The load on the contact is supported by an array of discrete and small microcontacts which, taken together, create a real contact area. The magnitude of the real contact area depends on the load, the surface roughness and mechanical properties of materials of the bodies in contact, and to a much lesser extent upon the gross contact geometry. On average, the real area of contact between engineering components is only 0.0001 to 0.01 of the nominal contact area. Profilometer-type instruments are used to measure this surface microtopography. The magnitude of the real contact area may also change with time. In the case of static friction, the size of the real contact area may increase with time due to creep. With kinetic friction there may be no change in the magnitude of the real contact area but there may be changes of a qualitative nature which reflect the modification of the surface topography and changes in the mechanical properties of the material resulting from frictional interactions. *Running-in* is where much of these occur.

Thus, the frictional interactions which occur during dry friction processes do not occur simultaneously over the whole nominal contact area but take place at discrete microcontact areas which are sometimes termed *elementary contact areas*. During the friction process, the elementary contact areas change in terms of their size, orientation, and location. The number of elementary contact areas per unit of the nominal contact area, their average size, location, form, and orientation constitute an important geometric characteristic of the friction process. This characteristic changes with time. Initially, the geometric characteristic is defined, to a considerable extent, by the surface roughness created during machining. At the later stages, the predominant influence on the geometric characteristic is introduced by the wear process. Moreover, the geometric characteristic of the friction process depends on the magnitude of the nominal contact stress and the tangential force as well as on mechanical properties of the materials in contact and debris incorporated from the environment.

From the energy point of view, friction is a process during which kinetic energy (or the work done by forces maintaining the relative motion of bodies in contact) is converted into other forms of energy. Usually the largest part of the mechanical energy lost during friction is converted into heat. In the early days of research on friction it was thought that all mechanical energy is converted into heat—actually, virtually all the energy is so degraded.

As a general guide, the following forms of energy can be distinguished into which the mechanical energy is converted during friction:

- Heat energy, causing an increase in temperature of rubbing bodies.
- Acoustic energy, producing audible effects.
- Optical energy, including the full spectral range.
- Electric energy, responsible for generating electrostatic charge.
- Mechanical energy, causing wear of contacting bodies.
- Mechanical energy (or entropy), causing further comminution of wear particles.

Specific forms of energy into which the mechanical energy lost during friction is converted and their contributions to the overall energy balance of the friction process depend on the nature of rubbing bodies and the contact conditions. It will be realized that eventually virtually all of the frictional work is degraded into heat. For this reason, it is observed in practice that there is little correlation between frictional work and wear.

The Dual Nature of Frictional Process

Fundamentally, in the context of historical precedence, at least two factors can be identified as being responsible for the energy losses generated during friction processes. One involves energy losses in relatively thin interface zones, and the other involves energy losses beneath the contact in much larger volumes of material.[1] The first factor is usually called the *adhesion* component of friction and the second the *deformation* component of friction. In the former, the strain energy is transmitted to the system through the natural adhesive forces which exist at the contacting asperities, hence the work varies to a first approximation with the real area of contact. This process was implicitly introduced previously. The deformation component does not rely, in the first-order approximation, upon surface forces but rather the localized geometry of deformation is responsible for the strain transmission and primary energy dissipation in the subsurface material. In the friction of metals where yield criteria are well established, the overall deformation is sliding is determined by the combined action of the surface traction and the subsurface bulk deformation. It has been shown[2] that when a hard wedge is slid over a softer, nonwork hardening metal the contribution from the two factors (adhesion and deformation) may be regarded as independent, to a good approximation, and thus may be added. In what follows, it shall be assumed that the friction process can be modeled as the sum of notional contributions from those two factors. Both of these components will be considered in more detail in later sections.

Phenomenology of Friction Process

Elementary friction phenomenon The idea of an elementary friction phenomenon is a basic notion and denotes everything which occurs within the microcontact, from the moment of the microcontact's creation to the moment of its destruction. All elementary friction phenomena which take place at a given moment within the nominal contact area constitute the physicochemical characteristics of the friction process. The physicochemical characteristics determine the magnitude of the frictional force and the type and intensity of the wear which results.

The primary basis for the development of the elementary friction phenomena is the mechanical interaction which occur between the asperities on the two contacting surfaces. This interaction is defined by the incremental normal force dN_i and tangential force dF_i which act on the microcontact. The third parameter determining the dynamic nature of frictional interaction is the relative velocity dv_i of two opposing asperities at the moment of contact. All three parameters change during the duration of the elementary friction phenomenon.

The state of stress that results in the element of surface from the mechanical interaction is very complex and variable. It is a three-dimensional state of stress depending upon, apart from the previously mentioned parameters dN_i, dF_i, and dv_i, the geometry of the microcontact, the geometry of interacting asperities, the material properties, the temperature, the frictional heating, and the adhesive interactions. All these factors are interdependent and changeable. The mechanical interactions produce, initially, elastic deformations of the surface elements which are usually followed by plastic deformations. Plastic deformations may, in turn, produce work hardening of the material, while the frictional heat can induce recrystallization, decrease the hardness, and enhance diffusion and chemical interactions between the material and the surroundings. Besides, the plastic deformations facilitate the creation of bonds between atoms and molecules and release elastic strains. Due to the complexity of the problem, its rigorous analytical treatment is practically impossible and therefore a simplified approach and compromise must be accepted.

In every elementary friction phenomenon for the following stages can be distinguished:

1 The establishment of the microcontact between surface asperities.

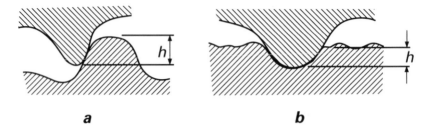

Figure 3.1 Establishment of a contact between two opposing asperities: a) geometrical interference and b) penetration

2 Physicochemical modifications of the microcontact and the surrounding material.

3 The breaking or rupture of the microcontact.

The contact between rubbing surfaces may have mechanical (deformation) or atomic (adhesive) nature, although simultaneous combination of both is also possible and indeed likely.

The establishment of the microcontact   The establishment of the microcontact can result from a mechanical interference of two opposing surface asperities or from penetration of a softer material by asperities of a harder one (see Figure 3.1). The ability of a hard asperity to penetrate the softer surface depends, to some extent, on its shape and the hardness ratio of the bodies in contact. An increase in the load on contact, surface finish, and the hardness difference facilitates the change from interference between surface asperities to penetration. In the case of very smooth surfaces, penetration is, practically, the only way of establishing the microcontact. In the majority of cases, however, the microcontacts are established as a result of interference and penetration taking place simultaneously.

When the normal approach of contacting bodies is sufficiently close, the atomic interactions become significant, creating adhesive junctions as well as supporting the normal load.

Physicochemical changes within the microcontact   Physicochemical changes within the microcontact are an inevitable result of maintaining the relative motion between contacting bodies. The following changes can be distinguished:

• Reversible and irreversible.

• Stabilized and progressive.

• Shallow and deep.

Reversible changes disappear immediately after their cause ceases to exist, while irreversible changes are retained permanently. Flash temperatures represent a case where no other changes are produced.

Stabilized changes represent finished processes which take place within the material and which do not progress any further although their cause might still be present. White layer formation in steels or cast iron is an example. Progressive changes develop as a result of repeated action of the same impulse.

Shallow changes are localized in the active uppermost layer of material. Deep changes affect those parts of a material which are beneath the active layer.

Breaking of microcontact The way in which the microcontact is broken depends on the history of the physicochemical changes which have taken place within the microcontact. Three basic forms of microcontact failure can be distinguished:

- Failure without any change in the initial geometry of contacting asperities.
- Failure involving change of shape of contacting asperities.
- Failure associated with the transfer of material from one asperity onto the other one.

The occurrence of any particular form of failure depends on the geometry of the contacting asperities, the magnitude of the interference or the depth of the penetration, and the mechanical properties of materials at the moment of microcontact failure. Figure 3.2 shows, schematically, a number of different ways in which a microcontact can fail.

Failure without change of the initial geometry of contacting asperities usually takes place during mechanical interactions leading to an elastic deformation, or during molecular or atomic interactions, provided that the failure occurs at the interface. The interference between the two opposing asperities can be modeled with the help of a cantilever beam analysis. Elastic penetration is controlled by the depth and the hardness of contacting materials and the contact geometry.

Failure at the interface usually involves the shearing of the adhesive bonds formed within the microcontact. For this form of failure to happen, the shear strength of adhesive bond should be less than the shear strength of both materials involved.

Failure involving a change of shape of the interacting asperities stipulates plastic deformations to take place within the microcontact. As a result of the interference, interacting asperities are flattened and the material is pushed in a direction which is opposite to that of the relative motion. During plastic penetration combined with relative motion, deformed material is pushed aside and the groove is formed on the surface of the softer material. Such a process is called *ploughing*. Failure associated with the transfer of material from one surface onto the other requires that the shear strength of adhesive junction is greater than that of the two materials involved. Failure of the microcontact then takes place within the weaker material, and as a result of that, transfer of material occurs. It is quite clear that the adhesive junctions are formed by molecular or atomic interactions between two bodies in close contact.

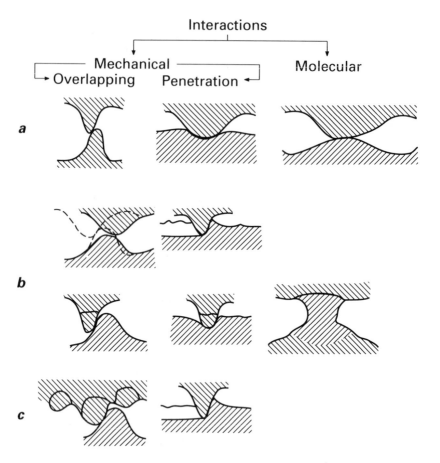

Figure 3.2 **Ways in which a contact between asperities can be terminated: a) without any change to the shape, b) with a change to the shape, and c) with creation of debris**

Real Area of Contact

Surface of a solid body With few exceptions, surfaces of solids are rough at a microscopic level. Even the surface of apparently smooth quartz is covered by 100 Å high asperities. Metallic surfaces with the best possible finish have asperities from 0.01 µm to 0.1 µm high. The roughness of a coarsely turned surface can be as much as 200 µm. The surface of a solid body is the boundary which separates that body from another body, substance, or space. It is defined by the mechanical forces emanating from the surface. The nominal surface is the intended surface contour, the shape and extent of which is usually shown and dimensioned on a drawing or descriptive specification. The measured surface is naturally a representation of the surface obtained by an instrument. In general, the properties abstracted will also be

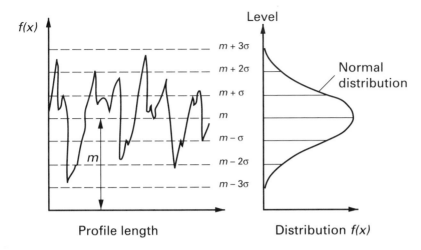

Figure 3.3 **Normal distribution of profile ordinates for machined surface**

a function of the instrument characteristics. An important set of parameters characterizing a solid body surface is called *texture*. Surface texture is the repetition of a random deviation from the nominal surface which forms the three-dimensional topography of the surface. The description of surface texture includes the specification of such parameters as the roughness, waviness, lay and flaws. Roughness consists of the finer irregularities of the surface texture, usually including those irregularities which result from the inherent action of the production process. These are considered to include traverse feed marks and other irregularities whose spacing is greater that the roughness sampling length and less that the waviness sampling length. Waviness may result from such factors as machine or work deflections, vibration, chatter, heat treatment, or warping strains. Roughness may be considered to be superposed on a wavy surface.

Lay is the direction of the predominant surface pattern, ordinarily determined by the production method used. Flaws are unintentional irregularities which occur at one place or at relatively infrequent or widely varying intervals on the surface. Flaws include such defects as cracks, blow holes, inclusions checks, ridges, scratches, etc. Normally, the effect of flaws is not included in the roughness average measurements.

The error of form is considered as being the deviation from the nominal surface which is not included in surface texture. Surface texture plays an important role in the friction process and therefore it must be measured and, if necessary, changed. There are a number of definitions and terms relating to the measurement of surface texture. The profile is the contour of the surface in a plane perpendicular to the surface. The measured profile is a representation of the profile obtained by an instrument (see Figure 3.3). When the measured profile is a graphical representation, it

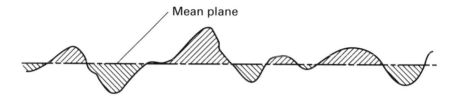

Mean plane

Figure 3.4   Centerline of the surface profile

will usually be distorted through the use of different vertical and horizontal magnifications but should otherwise be as faithful to the actual profile as technically possible. The graphical centerline is the line about which roughness is measured, and is a line parallel to the general direction of the profile within the limits of the sampling length, such that the sums of the areas contained between it and the parts of the profile which lie on either side are equal (see Figure 3.4). A peak is the point of maximum height on the portion of a profile which lies above the centerline and between the two intersections of the profile and the centerline. A valley is the point of maximum depth on the portion of a profile which lies below the centerline and between two intersections of the profile and the centerline.

The roughness average R_a, also known as the arithmetic average and the centerline average, is the arithmetic average of the absolute values of the measured profile height deviations taken within the sampling length and measured from the graphical centerline. The peak-to-valley height is the maximum deviation above the centerline plus the maximum deviation below the centerline within the sampling length. This value is typically three or more times the roughness average. The bearing area is a sum of all microcontact areas resulting from the contact between rough surface and an ideal smooth reference plane under given normal load. The magnitude of the normal load is such that only elastic deformations are produced. Figure 3.5 shows the geometry and parameters involved in the bearing area estimation. The reference plane is shown as a line displaced from its initial position by a distance a, which is the measure of a normal approach. The length of the bearing area L_i is the sum of all component lengths contained within the specified reference length L_p and resulting from the contact between individual asperities and the reference plane.

The linear share of the bearing area U_L is defined as the ratio of the length of the bearing area L_i to the reference length L_p:

$$U_L = \frac{L_i}{L_p} \tag{3.1}$$

where $L_i = b_1 + b_2 + b_3 + ...$

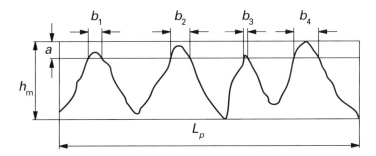

Figure 3.5 Definition of the length of the bearing area

Estimation of real contact area In principle, two rough surfaces will always make contact at a minimum of three points. When the normal load N is applied, the initial separation h between the surfaces decreases by the value of the normal approach a. As the normal approach increases, more and more surface asperities come into contact and undergo deformation. As a result, the number and the area of individual microcontacts increase, affecting the contact pressure distribution. Eventually, the real area of contact, i.e., the sum of all the microcontacts, reaches A_r and is sufficiently large to support the applied normal load N. Every microcontact, dA_{ri} is loaded by an elementary normal load dN_i. The overall normal approach a of the two contacting surfaces produces a decrease in the original height of the highest asperities by the amount:

$$a - [h - (h_{1i} + h_{2i})] \tag{3.2}$$

where h is the initial separation, and h_{1i} and h_{2i} are the original heights of two contacting asperities. Because of the differences in the decreases in the original heights as asperities and the local variation in the mechanical properties of the materials, individual asperities are in different states of stress, which results in differences in the magnitudes of deformation.

The magnitude of a normal approach a is a function of the normal load N, the nominal area of contact A_n, the surface roughness R_1 and R_2, and the mechanical properties W_1 and W_2 of the materials of contacting bodies. Thus,

$$a = f_1 (N, A_n, R_1, R_2, W_1, W_2) \tag{3.3}$$

The real area of contact is a function of a normal approach a:

$$A_r = f_2 (a) \tag{3.4}$$

Both functions f_1 and f_2 have a complex physical meaning and require certain assumptions regarding the conditions of the contact and the description of surface roughness in order to define their specific analytical forms.

Basically, two types of contact can be distinguished, namely those which exhibit elastic contacts and those which exhibit plastic contacts. It is usually assumed that the contact between two individual asperities has an inherent plastic nature, and therefore the real area of contact is given by

$$A_r = \frac{N}{\sigma_p} \tag{3.5}$$

where σ_p is the stress at which the softer material flows plastically. This stress can be estimated from the yield strength of the material:

$$\sigma_p = cQ_r \tag{3.6}$$

where c denotes a coefficient which depends on the shape and size of the ploughing asperity, and Q_r is a yield strength of the material. As a first approximation,[31] c can be taken as being equal to 3. The plastic flow stress is closely related to the Brinell hardness. Thus, the real area of contact is directly proportional to the normal load and inversely proportional to the plastic flow stress or hardness.

The case of an elastic contact is more complicated. The Hertz theory of elastic contact leads to the following formula for the real area of contact between a sphere and the flat surface:

$$A_r = N^{2/3}R^{2/3}\left[\frac{3E_a + 4G_a + 3E_b + 4G_b}{G_a(3E_a + G_a) + G_b(3E_b + G_b)}\right]^{2/3} \tag{3.7}$$

where R is the radius of the sphere, E is the Young's modulus, G is the modulus of rigidity, and a and b refer to the sphere and the flat surface respectively.

The main deficiency of the formula derived from Hertz theory is that surface roughness is not taken into account. However, there are methods which can be used to modify Hertz theory and to obtain a better estimate of the real area of contact. What is obvious is that tin the case of an elastic contact, the real area of contact increases with the load much more slowly than in the case of a plastic contact.

The general approach to the problem os sizing up the real area of contact is to use the bearing area curve together with the information regarding the character of asperity deformation.

Plasticity index   It is justified to say that when two rough surfaces are in contact there is an equal probability of finding a particular asperity height or finding any other height. Greenwood and Williamson[3] by assuming more plausible distributions of asperity heights showed that

$$\frac{N}{A_r} = \frac{E_{eq}}{k_D}\sqrt{\frac{\sigma}{r}} \qquad (3.8)$$

where E_{eq} is the equivalent elastic modulus given by:

$$\frac{1}{E_{eq}} = \frac{1}{2}\left(\frac{1-v_1^2}{E_1} + \frac{1-v_2^2}{E_2}\right) \qquad (3.9)$$

The quantity σ denotes standard deviation of asperity heights, r is an average radius of curvature at the peaks of asperities, v_1 and v_2 are the Poisson's ratios for the two surface materials, E_1 and E_2 are the respective Young's moduli, and k_D is a constant which depends on whether one assumes a Gaussian distribution of asperity heights (ϕ_g) or an exponential distribution (ϕ_e). These distributions are given by:

$$\phi_g = \left[\frac{1}{\sqrt{2\pi}}\right]\left(\frac{1}{\sigma}\right)\left[\exp\left(\frac{-z^2}{2\sigma^2}\right)\right] \qquad (3.10)$$

and

$$\phi_e = \left(\frac{1}{\sigma}\right)\left[\exp\left(\frac{-z}{\sigma}\right)\right] \qquad (3.11)$$

If it is taken into account that the real area of contact can be defined at the ratio of a normal load on the contact and the contact stress at which a softer material flows plastically, the following equation can be obtained:[3]

$$\left(\frac{E_{eq}}{k_D}\right)\sqrt{\frac{\sigma}{r}} = \sigma_p \qquad (3.12)$$

where usually $\sigma_p = 1.1\ Y$, and Y is the elastic limit of a material.

Since k_D is expected to be about unity, Greenwood and Williamson introduced the concept of *reduced flow pressure* given by:

$$\psi = \frac{E_{eq}}{\sigma_p}\sqrt{\frac{\sigma}{r}} \qquad (3.13)$$

When ψ, also called the *plasticity index*, is less than unity $E_{eq}\sqrt{\sigma/R}$ must be less than σ_p so that elastic deformation dominates. If $\psi < 0.6$ the probability of plastic flow is very small indeed. If ψ is greater than unity, a large part of the contact area will involve plastic flow. The plasticity index equation contains both elastic and plastic deformation parameters. The concept is not very different from the idea that plastic flow ensues once the mean pressure across the area of contact exceeds the flow pressure. However, ψ includes the statistical parameters describing the surface topography.

Adhesion Component of Friction

As was mentioned earlier, one of the most important components of friction originates from the formation and rupture of interfacial adhesive bonds. Extensive theoretical and experimental studies have been undertaken to explain the nature of adhesive interactions, especially in the case of clean metallic surfaces.[1] The main emphasis has been on the electronic structure of the bodies in frictional contact. Attractive forces within the contact zone include all those forces which contribute to the cohesive strength of a solid. An illustration of short-range forces in action is provided by two pieces of clean gold that have been pressed together and form metallic bonds over the regions of intimate contact. The interface will have a strength close to that of the bulk gold at the contact point. In contacts formed by polymers and elastomers, long-range van der Waals forces operate. Thus, the interfacial adhesion is as natural as the cohesion which determines the bulk strength of materials.

The adhesional component of the coefficient of friction is usually given, to first order, as the ration of the interfacial shear strength of adhesive junctions to the yield strength of the substrate material:

$$\mu_a = \frac{F_a}{N} \cong \frac{\tau_{1,2}}{\sigma_p} \tag{3.14}$$

where $\tau_{1,2}$ is the effective shear strength of the interface, and F_a is the frictional force resulting from shearing of adhesive junction. For most engineering materials this ratio is of the order of 0.2, which means that the friction coefficient may be of the same order of magnitude. In the case of clean metals, where junction growth is most likely to take place, the adhesion component of friction may increase to about 10 or even 100. The presence of any type of lubricant disrupting the formation of the adhesive junctions can dramatically reduce the magnitude of the adhesion component of friction. This simple model can be supplemented by the surface energy of the contacting bodies. The adhesional component of the friction coefficient is then given by[4]

$$\mu_a = \frac{\tau_{1,2}}{\sigma_p} \left[1 - 2 \frac{W_{1,2} \tan \theta}{\sigma_p} \right]^{-1} \tag{3.15}$$

where $W_{1,2} = \gamma_1 + \gamma_{1,2}$ is the interfacial surface energy.

Recent progress in fracture mechanics allows for fracture of an adhesive junction to be considered as a mode of energy dissipation. The formula for the adhesional component of the friction coefficient is given by[4]

$$\mu_a = C\frac{\sigma_{1,2}\delta_c}{n^2\sqrt{NH}} \tag{3.16}$$

where $\sigma_{1,2}$ is the interfacial tensile strength δ_c is the critical crack opening displacement, n is the work hardening factor, and H is the hardness of the substrate material.

It is important to remember that parameters such as the interfacial shear strength or the surface energy characterize a given pair of materials in contact rather than the single components involved. If, however, the interface zone is a well-defined material, say a solid lubricant, then the shear stress is a characteristic of the layer and the contact conditions; see below.

The Interface Shear Stress

The quantity τ, the interface shear stress, has been widely used as a means to prescribe the work done at interfaces during sliding. Its significance is still open to questions and the means whereby it has been measured leaves much to be desired. Typically, it is measured by shearing thin, usually lubricated layers in model contacts. The friction is measured and the real areas of contact is computed on the basis of various approximations.[32] The interface shear stress is found by experiment to be a function of all the contact variables, although the functional forms are not always accurately known. The most interesting variable is the mean contact pressure P, and

$$\tau = \tau_0 + \alpha P \tag{3.17}$$

where τ_0 is an intrinsic characteristic shear stress, and α is a pressure coefficient. Various values of τ_0 and α have been published. A little rearrangement gives

$$\mu = \frac{\tau_0}{P} + \alpha \tag{3.18}$$

When plastic flow occurs at asperities where the interposed film may be described by the τ_0 and α relationship we have

$$\mu = \frac{\tau_0}{H} + \alpha \tag{3.19}$$

where H is the hardness of the solid.

For many organic materials, α is on the order of 0.2. The relationship also shows that μ will lead to large values at low contact pressures, and this appears to be the case.

The functional dependence of τ on temperature and sliding velocity has been examined for certain systems. The interface shear stress normally decreases with the

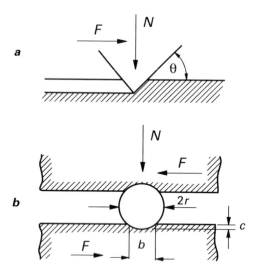

Figure 3.6   Ploughing component of friction

increasing temperature. The variation with sliding velocity, thermal heating effects apart, is more complex.[33–35]

The physical significance of τ continues to be discussed, and more is said about this in the conclusion. However, as far as can be deduced, it appears to reflect a shear process whose characteristics are very similar to those produced in a *bulk* shear process. The interface shear process is unique insofar as it corresponds to very high rates of strain and transient quasihydrostatic loadings. As such, these extremes are not reproduced in conventional bulk deformation processes. Hence, the correlation of interface shear processes and those in bulk shear is of necessity tentative in nature.

Deformation Component of Friction

Ploughing occurs when two bodies in contact have considerable different hardnesses. The asperities on the harder surface may penetrate the softer surfaces and if there is relative motion, produce grooves if the flow stress is exceeded. Because of the ploughing displacement, a certain force is required to maintain motion, and in certain circumstances, this force may constitute a major component of the overall frictional force. If a hard surface asperity is modeled as a cone making an angle θ with the surface (see Figure 3.6), then the ploughing component of the friction coefficient is given, in a first-order analysis, as[36]

$$\mu_p = \frac{2}{\pi}\tan\theta \tag{3.20}$$

for a rigid-plastic deformation.

As the asperities on engineering surfaces seldom have an effective slope greater than a few degrees, it follows, therefore, that the friction coefficient resulting from this formula should be on the order of 0.04. This is, of course, too low a value, mainly because the piling up of the material ahead of the moving asperity is neglected in this analysis.

Ploughing of a brittle material is inevitable associated with microcracking, and therefore a model of the ploughing process based on fracture mechanics is appropriate. Material properties such as fracture toughness, elastic modulus, and hardness are used then to estimate the ploughing component of the friction coefficient:

$$\mu_p = \frac{F_p}{N} \cong C\frac{K_{Ic}^2}{E\sqrt{HN}} \tag{3.21}$$

where K_{Ic} is the fracture toughness, E is the elastic modulus, and H is the hardness.

The problem of ploughing resulting from the presence of hard wear particles in the contact zone has received considerable attention because of its practical importance. It has been found that the frictional force produced by ploughing is very sensitive to the ratio of the radius of the particle to the depth of penetration. The formula for estimating the ploughing component of the friction coefficient in this case has the following form:

$$\mu_a = \frac{2}{\pi}\left[(\frac{2r}{b})^2 \sin^{-1}\frac{b}{2r} - \sqrt{(\frac{2r^2}{b})-1} \right] \tag{3.22}$$

Mechanical energy can also be dissipated through the deformations of contracting bodies during sliding where no groove is produced. The usual approach to the analysis of the deformation of a single surface asperity is the slip-line field theory for a rigid, perfectly plastic material. A slip-line deformation model of friction, shown in Figure 3.7, is based on a two-dimensional stress analysis. Three distinct regions of plastically deformed material may develop, and in Figure 3.7 they are denoted *ABE*, *BED*, and *BDC*. The flow shear stress of the material defines the maximum shear stress which can be developed in these regions. The deformation component of the friction coefficient is given by the expression[4]

$$\mu_d = \frac{F_d}{N} = \lambda \tan\left[\arcsin\left(\frac{\sqrt{2}\ (2+\beta)}{4\ (1+\beta)} \right) \right] \tag{3.23}$$

where $\lambda = \lambda\ (E, H)$ is the portion of plastically supported load, E is the elastic modulus, and H is the hardness. The proportion of load supported by the plastically deformed regions and related, in a rather complicated way, to the ratio of the hardness and the elastic modulus is an important parameter in this model. For

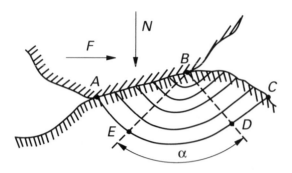

Figure 3.7   Deformation component of friction

completely plastic asperity contact and an asperity slope of 45°, the coefficient of friction due to deformation in 1.0. It decreases to 0.55 for an asperity slope approaching zero.

Another approach to this problem is to assume that the frictional work performed is equal to the work of the plastic deformation during steady-state sliding. This energy-based plastic deformation model of friction due to deformation gives the following expression for the coefficient of friction:[4]

$$\mu_d = \frac{A_r}{N} \tau_{max} f\left(\frac{\tau_{1,2}}{\tau_{max}}\right) \tag{3.24}$$

where

$$f\left(\frac{\tau_{1,2}}{\tau_{max}}\right) = 1 - 2 \frac{ln\left(1 + \left(\frac{\tau_{1,2}}{\tau_{max}}\right) - \left(\frac{\tau_{1,2}}{\tau_{max}}\right)\right)}{ln\left[1 - \left(\frac{\tau_{1,2}}{\tau_{max}}\right)^2\right]} \tag{3.25}$$

In this expression A_r is the real contact area, τ_{max} denotes the ultimate shear strength of a material and $\tau_{1,2}$ is the mean interfacial shear strength.

Viscoelastic Component Friction

The two-component model of friction presented earlier is mainly applicable to metals. In the case of elastomers, it is proper to talk about the adhesion term of friction F_a, and the hysteresis term of friction F_{hyst}. Thus,

$$F = F_a + F_{hyst} \tag{3.26}$$

Unlike a hard material, an elastomer structure is composed of flexible chains which are in a constant state of thermal motion, and therefore adhesion at a given velocity and temperature range is described by a thermally activated viscoelastic loss process.

During relative sliding between an elastomer and a hard surface, the separate chains in the surface layer attempt to link with molecules in the hard base, thus forming local adhesive junctions. Motion causes these bonds to stretch, rupture, and relax before new bonds are made, so that effectively elastomer molecules jump a molecular distance to their new equilibrium position. Thus, a dissipative stick–slip process on a molecular level is fundamentally responsible for adhesion, and several theories exists to explain this phenomenon in more detail;[35] see also the discussion of Schallamach waves in a later section.

From the macroscopic viewpoint, we can write for the total adhesional force[37]

$$F_a = \sum_{i=1}^{m} F_i = \sum_{i=1}^{m} A_i s_i \tag{3.27}$$

Where s_i is the local effective shear strength of the interface. On a molecular level, we can write at each location i

$$F_i = n_i j_i \tag{3.28}$$

where n_i is the number of molecular junctions or bonds between the elastomer and base at location i, and j_i is the local effective junction strength. From the last two equations, it follows that

$$s_i = \left(\frac{n_i}{A_i} \right) j_i \tag{3.29}$$

so that the local shear strength of the macroscopic model is a function of the molecular junction strength j_i. The coefficient of adhesional friction is

$$\mu_a = \frac{F_a}{N} = \sum_{i=1}^{m} \frac{n_i j_i}{N} \tag{3.30}$$

The various molecular kinetic stick-slip theories of adhesion for elastomers essentially attempt to establish how the number of macroscopic sites

$$M = \sum_{i=1}^{m} n_i \tag{3.31}$$

and j vary with loading, speed, temperature, and lubrication. On the other hand, mechanical model theories examine in a less precise but simplified manner the variation of contact area A_i and s with the same parameters.

The hysteresis component of friction results from the asymmetry of pressure distribution in the contact between a hard surface asperity and an elastomer.

In the absence of relative motion, a symmetrical draping of the elastomer material about the contour of the asperity is produced. If, however, the elastomer is moving with a finite velocity V relative to the base surface, it tends to accumulate or "pile up" at the leading edge of the asperity and to break contact at a higher point on the downward slope. Thus the contact arc moves forward compared with the static case. This effect creates an asymmetric pressure distribution where the horizontal pressure components give rise to a net force F_{hyst}, which opposes the sliding motion.

The hysteresis component can be viewed as the *inertia* associated with the portion of draped elastomer between asperities at any instant during the sliding process. This inertia causes a delayed reaction within the bulk of the elastomer, which causes materials to accumulate on the positive flanks of asperities and to fail to follow the contour of the latter on the negative slopes.

The coefficient of hysteresis friction is given by[37]

$$\mu_{hyst} = \frac{F_{hyst}}{N} \tag{3.32}$$

If other parameters characterizing the contact are taken into account, the following expression for the coefficient of hysteresis is obtained:

$$\mu_{hyst} = \left(\frac{4\gamma}{\lambda^3}\right)\left(\frac{E_d}{p}\right) \tag{3.33}$$

where γ is an asperity density factor accounting for the degree of *packing* of the asperities on the surface (usually, $\gamma \leq 1$), λ is the mean wavelength of the surface topography, E_d represents the energy dissipated per asperity contact, and p is the contact pressure.

Friction under Boundary Lubrication Conditions

When two metallic surface s in relative motion are separated by a lubricating film whose thickness is greater that the combined surface roughness, then the frictional force is mainly the result of viscous energy dissipation. Where exclusive subsurface deformations occur, for example in the case of elastomers, these dissipation processes, as shown earlier, also contribute to the frictional work. This is the well-known case of hydrodynamic lubrication. However, there are many cases where, for various reasons, this type of lubrication is impossible to achieve. Then, the thickness of a lubricating film is less than that needed to completely separate the contacting surfaces, and the magnitude of frictional force depends primarily on physicochemical characteristics of the lubricant and the metal surfaces. This type of lubrication is referred to as boundary lubrication.

Fractional film defect concept   Bowden et al.[5] proposed the following expression to describe friction during boundary lubrication:

$$F = \beta A_r \tau_m + (1 - \beta) A_r \tau_l \qquad (3.34)$$

where F is the frictional force, A_r is the real area of contact, τ_m is the shear strength of the adhesion junction, and τ_l is the shear strength of the boundary film. The fractional film defect β is given by the ratio

$$\beta = \frac{A_m}{A} \qquad (3.35)$$

where A_m is the metal–metal area of contact.

Taking into account these expressions, the frictional force can be written as

$$F = A_m \tau_m + (A_r - A_m) \tau_l \qquad (3.36)$$

but

$$A_r = \frac{n}{\sigma_p} \text{ and } N = A_r \sigma_p \qquad (3.37)$$

Therefore, the coefficient of friction under boundary lubrication is

$$\mu = \beta \frac{\tau_m}{\sigma_p} + \left[\frac{\tau_l}{\sigma_p} - \beta \frac{\tau_l}{\sigma_p} \right] \qquad (3.38)$$

If contributions from the adhesive component of friction and the viscous component of friction are taken into account, that is,

$$\mu_a = \frac{\tau_m}{\sigma_p} \text{ and } \mu_l = \frac{\tau_l}{\sigma_p} \qquad (3.39)$$

then the friction coefficient under boundary lubrication is

$$\mu = \beta \mu_a + (1 - \beta) \mu_l \qquad (3.40)$$

Relationship between friction and lubricant adsorption   Lubricating oils are normally formulated to include relatively small quantities of surface-active additives which adsorb, perhaps reversibly, onto the sliding surfaces. It is often supposed that this process may be treated using equilibrium chemical thermodynamics.

At low sliding velocities, the adsorbed molecules of a lubricant, outside the contact zone, have ample time to desorb, thus permitting metal–metal contact to occur. At high velocities, the time is insufficient for desorption, and metal–metal contact will be prevented. Kingsbury[6] proposed a relationship between the fractional film defect and the ratio of the time t_x for the surface asperity to travel a distance equivalent to the diameter of the adsorbed molecule and the average time t_r that a molecule remains at a given surface site. Using this concept, the fractional film defect can be described as[7]

$$\beta = 1 - \exp\left(-\frac{t_x}{t_r}\right) \tag{3.41}$$

The average time t_r spent by the molecule at the same site is given by[8]

$$t_r = t_0 \exp\left(\frac{E_c}{RT}\right) \tag{3.42}$$

where E_c is the energy required to bring about the desorption of the molecules and can be regarded as being equivalent to the isosteric heat of adsorption, R is the universal gas constant, T is the absolute temperature in the contact zone, and t_0 may be considered to a first approximation as a characteristic period of vibration of the adsorbed molecule.

If this is taken into account, the fraction film defect can be expressed as,

$$\beta = 1 - \exp\left\{-\left[\frac{30.9\times10^5\, T_m^{1/2}}{VM^{1/2}}\right]\exp\left(-\frac{E_c}{RT}\right)\right\} \tag{3.43}$$

where M is the molecular weight of the lubricant, and T_m is the melting temperature of the lubricant. Values of T_m are readily available for pure compounds only. In Equation 3.43, sliding velocity is denoted by V. Lubrication failure may then be supposed to occur at some critical temperature or sliding velocity.

The model has its limitations, not the least of which is the assumption of isothermal contact conditions. Flash temperatures will occur in the contact zones for example; see the following section. An alternative approach which leads to basically the same result has been offered by Cameron and his associates.[38] Again, a fractional film defect notion is assumed at the initiation of scuffing. However, a conventional thermodynamic analysis is developed which assumes a reversible adsorption process.[33] More is said about scuffing, or frictional catastrophe, in the next section.

Phenomena Associated with Friction

Scuffing as an uncontrolled friction process   Scuffing is usually defined as a gross damage characterized by the progressive formation of adhesive junctions.[9] For metallic surfaces to weld together, the intervening films on at least one of them must become disrupted and metal must move through the disrupted film to adhere to the other surface. There is an appreciable range of loads over which plastic flow takes place beneath the surface without it extending to the surface layers themselves. Under these conductions welding does not occur. However, when sliding is introduced, a tangential stress field due to friction is added to the normal load stresses. As it increases, the region of maximum shear stress moves upward while simultaneously a second region of high yield stress develops on the surface behind the circle of contact. The shear stress at the surface is sufficient to cause flow when the coefficient of friction reaches about 0.27. With plastic deformation in the surface layer itself, welding becomes possible.

Undoubtedly, scuffing is a complex phenomenon controlled by a number of factors such as load, friction, lubrication, speed of relative motion, surface topography, and surface temperature in the contact region. Recognizing the complexity of the problem, it has been postulated that scuffing is best considered in probabilistic terms.

A number of models and scuffing hypotheses have been put forward,[10, 11] and all of then treat the problem in deterministic terms. Blok[10] advanced the hypothesis that the factor controlling the breakdown of boundary lubricant film is the total temperature in the contact, which consists of two components. The first component is the bulk temperature of the metal parts at positions remote from the actual contact points. The second contribution is the instantaneous temperature rise as the surface passes through the heat source represented by the contact zones. Blok's postulate is that the boundary film will break down and scuffing will be initiated if the total contact temperature exceeds a critical value. As the critical temperature depends on a number of parameters characterizing the materials in contact and the lubricating medium, it is quite obvious that it cannot be predicted in a deterministic way but rather a probability that it will assume a certain value.

The central premise in the model advanced by Dyson et al.[11] is that scuffing can be avoided if a very thin film of lubricant of very high viscosity is present between pairs of interacting surface asperities. However, this high viscosity cannot be provided effectively by the local microelastohydrodynamic action of contacting asperities but only by the high pressure produced by operation of the main macroelastohydrodynamic system. This model of scuffing is deterministic and does not take into account the inherent variability of all the parameters involved.

Stolarski[12] has proposed a model of scuffing based on a probabilistic approach. In this model, the pr0obability of scuffing is mainly influenced by two parameters: the plasticity index discussed earlier and the temperature in the contact. The necessary condition for scuffing to be a serious prospect is when the plasticity index $\psi > 0.6$, which means the development of a plastic asperity contact. The sufficient condition is when $T > T_{cr}$, that is, temperature in the contact is greater than a critical temperature defined mainly by the heat of adsorption of a boundary lubricant and estimated on the basis of the fractional film defect presented earlier. When both conditions are simultaneously fulfilled, then

$$P \text{ (plastic asperity contact)} < P \text{ (scuffing)} < P \text{ (asperity contact)} \tag{3.44}$$

where P denotes probability of the event in the brackets.

The probability of plastic asperity contact is given by the expression[12]

$$P(\Delta h' \leq 0) = \frac{1}{\sqrt{2\pi}\sigma_*} \int_0^\infty \exp\left[\frac{-(\bar{h} + \delta_p - \Delta h')}{2\sigma_*^2}\right] d(-\Delta h') \tag{3.45}$$

The probability $P(\Delta h' \leq 0)$ of asperity plastic contact can be found from the normalized contact parameter $\Delta \bar{h}'$, where $\Delta \bar{h}' = (\Delta \bar{h} + \delta_p)/\sigma_*$ is the number of standard deviations from the mean value of separation h between contacting surfaces. The probability of asperity contact is given by,[12]

$$P(\Delta h \leq 0) = \frac{1}{\sqrt{2\pi}\sigma_*} \int_0^\infty \exp\left[\frac{-(\bar{h} - \Delta h)^2}{2\sigma_*^2}\right] d(-\Delta h) \tag{3.46}$$

In the foregoing, $\bar{h}$ is the mean value of the separation between contacting surfaces and σ_* is the resulting standard deviation of peaks on both contacting surfaces. Again, the probability $P(\Delta h \leq 0)$ of asperity contact can be found from the normalized contact parameter Δh, where $\Delta \bar{h} = \bar{h}/\sigma$ is the number of standard deviations from the mean h.

The probability of scuffing is determined in an indirect way. Its magnitude is not represented by a precise number but is given in the form of a range of values defined by the probability of plastic asperity contact.

Friction-induced relaxation vibration It is generally observed that the sliding of a body at slow velocities under dry or boundary friction is not a smooth continuous process. The motion of a body proceeds with jerks or oscillations. Contacting surfaces stick together until, as a result of a gradually increasing shear stress, there is a sudden break with a consequent very rapid slip. The surfaces stick again and the process is repeated indefinitely. This intermittent motion consisting of alternate clutching and breaking away of sliding surfaces is known as *stick–slip sliding*.

The phenomenon is observed in many practical applications, such as members sliding at low velocities in machine tools, mechanisms, and measuring apparatus. In many precision applications such as intermittent motion, apart from causing increased wear and undesirable stresses in the sliding members, affects adversely the accuracy an sensitivity of a shifting or a movement.

The phenomenon of intermittent motion has been subject of many studies. Wells[13] in 1929 observed stick–slip while trying to measure kinetic friction coefficient at low speeds. The vibration associated with stick–slip motion was later studied by Thomas,[14] who used graphical and analytical techniques to solve the differential equations of motion. Kaidanovsky and Haiken[15] noted the existence of a negative slope region in the friction–velocity curve for rubbing surfaces undergoing friction-induced vibration. Blok[16] linearized the friction–velocity curve and treated the problem analytically. Rabinowicz[17] studied the stick–slip process experimentally and obtained curves of amplitude of vibration versus driven surface velocity, but no theoretical argument was given as to their shape. Derjagin, Push, and Tolstoi[18] assumed a constant value of static friction which was independent of time of stick of the surfaces. Brockley et al.[19] carried out a theoretical analysis of friction-induced vibration and showed the existence of a critical velocity which limits the incidence of vibration. Hirst[20] showed that the stick–slip induced vibration depends on the overall stiffness of the system.

If a tangential force F, whose magnitude is less than the static friction force required to initiate a macroscale sliding, is applied to one of contacting bodies it will cause a certain definite displacement of the body to its new equilibrium position. This phenomenon, which is known as a junction growth or *primary displacement*, was first investigated[21] by Wierchovsky in 1926. If the force F is further gradually increased it will eventually attain a magnitude at which macroscopic scale motion commences. The condition for that to happen is

$$\frac{dF}{d(\Delta x)} = 0 \tag{3.47}$$

where Δx denotes the magnitude of primary displacement. The phenomenon of primary displacement has been subject of a number of studies.[22–25] The core of the problem was to find out whether the primary displacement is a result of elastic or plastic deformations. It was found experimentally[21] that the primary displacement consists of two components, that are elastic and plastic in type.

There is also the complication that relative sliding motion of contacting bodies does not always take place at constant velocity. Depending on the contact conditions and the elastic characteristic of the system, the velocity might be changing in a pulsating way leading to relaxation vibration of the system. It is now known that the relaxation vibration occurs in elastic systems when the static coefficient of friction is greater than the kinetic coefficient of friction. An elastic system can be represented

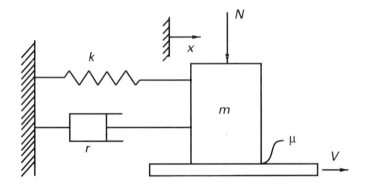

Figure 3.8   Sliding system

in the way depicted in Figure 3.8. A slider of mass m is in contact with a lower surface. A normal force N is applied prior to the commencement of relative motion between the slider and the lower surface. The usual assumption is that the movement of the slider is restrained by spring of stiffness k and a damper with damping coefficient s. The absolute displacement x is measured from the unstrained equilibrium position of the spring. If the coefficient of friction μ between the lower surface and the slider is sufficiently large at the equilibrium position, the slider will stick to the lower surface and move along with it at velocity V. The relationship between forces acting on the system during the stick period is

$$sV + kx < \mu_k N \qquad (3.48)$$

where μ_s is the static coefficient of friction. The subsequent slip period can be described by the equation

$$m\ddot{x} + s\dot{x} + kx = \mu_k N \qquad (3.49)$$

where μ_k is the kinetic coefficient of friction and may vary with velocity. It must, in fact, decrease from the static value if relaxation vibrations are to occur. The equation governing the motion of the slider during slip is a second-order differential equation and has been solved by a number of authors.[19] What is of most interest to an engineer is to find out how the stick–slip motion can be eliminated. First, it is necessary to note that the trouble stems from the interaction of the driving system dynamics with the friction characteristic of the loaf supporting surfaces. One approach relevant in the context of this chapter is to make the frictional forces so small that their variation with velocity is of no importance. This may be achieved by using hydrostatic bearings, or surfaces coated with low-friction material, or by changing from sliding friction to rolling friction, which is discussed next.

3.3 Rolling Friction

Review of Rolling Friction Hypotheses

The first experimental research on sliding friction was carried out by Leonardo da Vinci in 1508 and the concept of the coefficient of rolling friction arose as a result of these experiments.

In early research, rolling friction was treated as a mechanical process; that is, the interaction of rough surfaces of absolutely rigid solid bodies was considered.

The concept of rolling friction as a mechanical process was introduced by Delagir and was further developed and studied by Parin, Euler, Lesley, and others.

Coulomb was the first to introduce the hypothesis of the dynamic nature of friction during rolling.[26] The experimentally obtained relationship of the frictional force due to rolling along a place is given by the classical equation

$$F = \mu_r \frac{N}{r} = \mu N \tag{3.50}$$

where μ_r is the coefficient of rolling friction, having the dimension of length in this expression; N is the normal load; r is the radius of the rolling body; and μ is a coefficient similar to the coefficient of sliding friction.

In 1876 Reynolds[27] proposed the hypothesis of relative slipping according to which the frictional force due to rolling a perfectly elastic body along a perfectly elastic plane appears to be the result of relative slipping of contact surfaces due to their deformation. It can be seen from Figure 3.9 that in the region AC surface layers of the roller are compressed in the direction along the area of contact and are elongated in the plane. In the region CB of the contact area these deformations take place in the opposite direction, as a result of which microslip occurs.

Tomlinson[23] and later Tabor[28] showed that slipping in the contact zone in not the only source of frictional loss in rolling. In this connection, the hypothesis of molecular interaction was put forward. It was shown experimentally[28] that even during the rolling of two cylinders or two balls of identical size made of identical material, loss of energy occurs. According to Tomlinson, when two bodies approach each other, atomic and molecular interactions appear and energy is lost in overcoming them, which accounts for the additional energy loss.

Taking into account the forces of molecular interaction during the rolling of a cylinder over a plane, Tomlinson determined the frictional coefficient as follows:

$$\mu_r = \frac{3}{4} \sqrt{\frac{\pi}{2} \frac{e\mu}{\sqrt{Nr\theta}}} \tag{3.51}$$

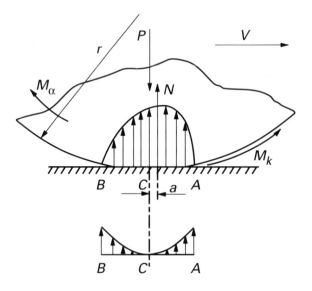

Figure 3.9   Graphical illustration of Reynold's hypothesis

where μ_r is the coefficient of rolling friction, N is the load on contact, r is the radius of the cylinder, e is the crystal lattice constant, μ is the coefficient of sliding friction, and θ is a function of the elastic constants of the materials of the bodies in contact.

Tomlinson carried out experimental verification of the coefficient of rolling friction by introducing an additional coefficient λ related to the coefficient of rolling friction μ_r by the expression $K = \mu l$, where l is the moment arm applied to the rolling body. The coefficient λ was determined by the damping of the oscillations of a pendulum. The value of λ thus obtained was twice the theoretical value for amplitudes of pendulum oscillation ranging from 4° to 0.61°, and this was explained by the fact that the relative displacement and molecular sizes are commensurable.

Ishlinskii[29, 30] and Tabor[28] studied the relationship between the frictional losses during rolling and imperfect elastic properties of true solids. Consider a perfectly hard roller moving over a viscoelastic surface and over a yielding surface. The forces acting on the roller are shown in Figure 3.10. The presence of a frictional force in this case is explained by the asymmetric contact stress distribution.

Ishlinskii[30] developed a relationship for the determination of the rolling resistance force F (at small rolling speed) in the case of a perfectly hard cylinder rolling over a yielding surface:

$$F = \frac{\mu_i v}{Cr} N \qquad (3.52)$$

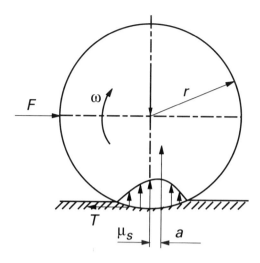

Figure 3.10 Appearance of rolling friction force with asymmetric distribution of contact pressure

where N is the load on the cylinder, v is rolling velocity, μ_i coefficient of internal friction of the substrate material, r is the radius of cylinder, and C is the hardness coefficient of substrate material.

The coefficient of rolling friction is thus obtained as

$$\mu_r = \frac{\mu_i v}{C} \tag{3.53}$$

Free Rolling

This represents the most basic form of rolling motion, and is best represented by the case of a cylinder or ball rolling without constraint in a straight line along a plane.

During rolling motion material ahead of the contact region is compressed, while material to the rear of the contact zone has this stress state relieved. For these conditions Hertz theory applies. Two cases are going to be presented here, a cylinder on a plane and a sphere on a plane.

Cylinder on a plane The distribution of pressure p in the contact zone due to an applied load N is

$$p = \frac{2N}{\pi a l}\sqrt{1 - \frac{x^2}{a^2}} \tag{3.54}$$

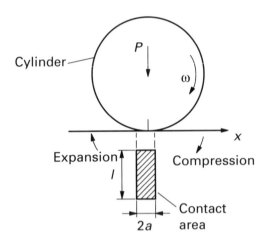

Figure 3.11 Contact between a cylinder and a plane

This equation represents a semicircular pressure distribution with a maximum pressure at the center of the contact zone of value $2N/(\pi al)$ as shown in Figure 3.11. The value of the contact semiwidth a is given by

$$a = \sqrt{\frac{4Nr}{\pi l}\left(\frac{1-v_1^2}{E_1} + \frac{1-v_2^2}{E_2}\right)} \qquad (3.55)$$

where v is Poisson's ratio, E is Young's modulus, and the suffixes 1 and 2 denote the bodies shown in Figure 3.11.

During the forward compression of the material in contacting bodies a resisting moment M about the center of the contact region is developed:

$$M = \int_0^a plxdx = \frac{2Na}{3\pi} \qquad (3.56)$$

Under ideal conditions the cylinder would be subjected to an equal and opposite moment arising from the pressure distribution in the rear part of the contact zone, but the rearward recovery does not replace the whole of this elastic work of forward compression, which is given by

$$\phi = \frac{Mx}{r} = \frac{2Nax}{3\pi r} \qquad (3.57)$$

This imbalance may be explained by an energy dissipation due to the elastic hysteresis loss occurring in the complex straining of the material which must occur during the rolling process. If this loss is represented by a coefficient ε, then

$$Fx = \varepsilon \phi = \varepsilon \frac{2Nax}{3\pi r} \tag{3.58}$$

where F is the force required to overcome the resistance to motion. The ratio F/N is usually called the coefficient of rolling friction μ_r,

$$\mu_r = \frac{F}{N} = \frac{2}{3}\frac{\varepsilon a}{\pi r} \tag{3.59}$$

It must be noted that μ_r depends on the geometry of the roller, the applied load, and the elastic constant of the two bodies in contact. Moreover, the hysteresis loss factor ε used in the expression for μ_r is not the same as that which would be measured in the simple tensile test. An approximate estimate of the hysteresis loss coefficient suggests that it should be about three times the loss factor measured in a simple tension test.

Sphere on a plane   The pressure distribution in the contact zone is given by

$$p = \frac{3N}{2\pi a^2}\sqrt{1 - \frac{r^2}{a^2}} \tag{3.60}$$

and the radius of the contact by

$$a = \left[\frac{3}{4}Nr\left(\frac{1 - \vartheta_1^2}{E_1} + \frac{1 - \vartheta_2^2}{E_2}\right)\right]^{1/3} \tag{3.61}$$

The contact configuration of the case under consideration is shown in Figure 3.12. The moment resisting motion is

$$M = \frac{3N}{4a^3}\int_0^a (a^2 - x^2)\,x\,dx = \frac{3Na}{16} \tag{3.62}$$

The work done in forward compression in rolling a distance x is

$$\phi = \frac{Mx}{r} = \frac{3Nax}{16r} \tag{3.63}$$

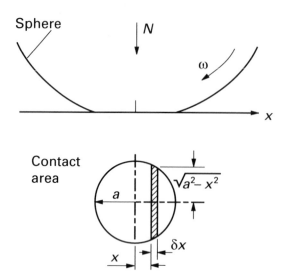

Figure 3.12 Contact between sphere and a plane

and introducing a loss factor similar to the one discussed earlier

$$Fx = \varepsilon\phi = \varepsilon\frac{3\,Nax}{16\,r} \tag{3.64}$$

Therefore, the coefficient of rolling friction is

$$\mu_r = \frac{F}{N} = \frac{3}{16}\frac{\varepsilon a}{r} \tag{3.65}$$

The hysteresis loss coefficient for this case is only about twice that which will be obtained in a simple tension test.

3.4 Exceptional Friction Processes

So far we have dealt with the two classical frictional components almost entirely from the stance of continuum mechanics. The microscopic facets are introduced via the discrete activate real areas of contact and dry discrete asperity deformations.

There are two facets of friction which rather fall outside this theme. They are Coulombic friction processes and the phenomenon of Schallamach friction. Neither is commonly encountered in engineering practice.

Coulombic friction is an old idea, and it is well recorded in Dowson's book.[39] It is sometimes referred to as "ratchet friction." Interlocking asperities must climb up and over each other, and by this means energy is dissipated outside the contact zones. There is no plastic or viscoelastic loss in the contact zones. Although the mechanism is now largely discredited, recently it has enjoyed new favor in the context of powder and fiber friction processes.[40]

Schallamach friction is a process which has been established as a credible means of energy dissipation for elastomers sliding over smooth clean substrates. Sliding does not occur at the interface, but instead a macroscopic dislocation propagates through the contact. Frictional energy is dissipated by the forming and peeling apart of this ruck as it passes across the zone of contacts. In essence, it is a special type of mode 1 fracture or adhesion experiment. The process seems to be restricted to systems which can reasonably accommodate a strain of at least unity.[41]

At this point, it is necessary to make a brief note of the new fashion in microscopic friction modeling which may involve computer simulations. A recent NATO conference on friction[42] contains some discussion on this topic, as well as conventional frictional models or interpretations of friction. Recently, we have surveyed this topic.[34] Microscopic or molecular models have a long history. In their current form, as well as historically, they have been microscopic Coulombic models involving atomic or atomic group "asperities." They have been recently called "cobblestone" models, but like their historical precedents, they are inherently nondissipative. As such, they are certainly not viable for describing kinetic friction but they may have merits for static friction models. Essentially, they are simply models for the theoretical shear yield stresses for solids. The requirement for a microscopic deformation model is to specify a realistic and quantitative means of dissipating energy, and attempts are being made to produce the same. At least the problem of energy dissipation is now being directly addressed.

The main problem in specifying friction is possibly elsewhere, however. It is likely that, as with bulk solids, defects, and in particular, dislocation processes are involved in an interplane whose structure and properties have evolved as a consequence of the frictional work induced during sliding. Much needs to be done to accurately describe this plane or third body in an appropriate way.

3.5 Recapitulation

The study of friction has a long and distinguished history. The history has been one of an attempt to identify groups of mechanisms based largely upon the scales of deformation within the interface energy dissipation zones. The adhesion model focuses upon the surfaces near the interface and prescribes a thin interface work zone. The real contact area is important, and a conventional interface shear stress is used to define the specific energy dissipation rates. It is a good model as far as it goes. The deformation models involve a larger scale of deformation and energy loss. They are plausible and relate to bulk deformation studies. Coulombic or ratchet friction is a special but interesting case which has been rediscovered in the context of small contact pairs such as powders and fibers. Interestingly, the model is the first basis for the classic and now rediscovered molecular (or atomic) models which enjoy favor among those who seek to develop computer simulations of frictional processes. The real challenge is the atomic or molecular basis of friction which will combine all the traditional approximations. This may be a long time coming. In the meanwhile we may seek comfort in the predictive value and physical insights provided by the adhesion and deformation models as embodied in the two-term noninteracting model of friction.

References

1 F.G. Bowden and D. Tabor, *Friction and Lubrication of Solids,* I and II, Oxford University Press, 1950 and 1963.

2 J.M. Challen, L.J. McLean, and P.L.B. Oxley, *Proc. R. Soc. London,* **A394,** 161, 1984.

3 J.A. Greenwood and J.B.P. Williamson, Contact of nominally flat surfaces, *Proc. R. Soc. London,* **A295,** 300, 1966.

4 T.A. Stolarski, *Tribology in Machine Design,* Heinemann Newnes, 1990.

5 F.P. Bowden, J.N. Gregory, and D. Tabor, Lubrication of metal surfaces by fatty acids, *Nature,* **156,** 279, 1945.

6 E.P. Kingbury, Some aspects of the thermal desorption of a boundary lubricant, *J. Appl. Phys.,* **29,** 6, 1958.

7 T.A. Stolarski, Evaluation of the heat of adsorption of a boundary lubricant, *ASLE Trans.,* **30,** 472, 1987.

8 F.A. Lindemann, The calculation of molecular vibration frequency, *Physik Zeitung,* **11,** 20, 1910.

9 W. Hirst, Scuffing and its prevention, *Chartered Mech. Engr.,* April, 188, 1974.

10 H. Blok, The postulate about the constancy of scoring temperature, in *Interdisciplinary Approach to the Lubrication of Concentrated Contacts,* NASA SP-237, 1970.

11 R.W. Snidle, S.D. Rossides, and A. Dyson, The failure of end lubrication, *Proc. R. Soc. London,* **A395,** 291, 1984.

12 T.A. Stolarski, Probability of scuffing in lubricated contacts, *J. Mech. Eng. Sci.,* **203,** 361, 1989.

13 J.H. Wells, Kinetic boundary friction, *The Engineer, London*, 147, 54, 1929.

14 S. Thomas, Vibrations damped by solid friction, *Phil. Mag.*, 9, 329, 1930.

15 N.L. Kajdanovsky, S.E. Haiken, Miechanitcheskije relaksacyonnoye kolebanya, *Zhurnal Technitcheskoy Fiziki*, 3 (1), 1933.

16 H. Blok, Fundamental aspects of boundary friction, *J. Soc. Automotive Eng.*, 46, 275, 1940.

17 E. Rabinowicz, *Friction and Wear of Materials*, John Wiley and Sons, 1965.

18 B.C. Derjagin, V.E. Push, and D.M. Tolstoi, A theory of stick-slip sliding of solids, in *Proc. Conf. Lubrication and Wear*, 1957.

19 C.A. Brockley, R. Cameron, and A.F. Potter, Friction-induced vibration, *Trans. ASME J. Lub Techn.*, 101, 1967.

20 J.R. Bristow, *Proc. R. Soc.*, A189, 88, 1947.

21 A.W. Wierchovsky, Yavlenije pridvaritielnego smiestchenija pri trogani niesmazanyh povierhnostiej, *Zhurnal Prikladnoy Fiziki*, II (3–4), 1926.

22 I.S. Rankin, The elastic range of friction, *Phil. Mag.*, 8 (2), 1926.

23 G.A. Tomlinson, Molecular theory of friction, *Phil. Mag.*, 11 (7), 1929.

24 I.V. Kragelskii, Trenye voloknistyh viestchestv, Gizleprom, Moskva, 1941.

25 S.E. Haiken, A.E. Salomonovitch, and L.P. Lisovsky, I silah suhegho trenya, *Izdat. AN SSSR*, (1), Moskva, 1939.

26 C.A. Coulomb, *Theorie des machines simples eu ayont eu regard au frottment de laures parties et a la rainder das cardages*, Paris, 1809.

27 O. Reynolds, On rolling friction, *Phil. Trans. R. Soc.*, 166, 1876.

28 D. Tabor, The mechanism of rolling friction, *Proc. R. Soc. London*, A229, 198, 1955.

29 A.Yu. Ishlinskii, *Prikladnya Matematika i Mekhanika*, 2 (2), 1938.

30 A.Yu. Ishlinskii, Trenie i iznos v masinakh, Trudy Vsesoyuznoi Konferentsii, 2, *Izd. AN SSSR*, Moskva, 1940.

31 D. Tabor, *Hardness of Metals*, Oxford Univ. Press, Oxford, 1951.

32 B.J. Briscoe and D. Tabor, Shear properties of thin polymeric films, *J. Adhesion*, 9, 145, 1978.

33 B.J. Briscoe and D. Tavor, Lubrication, *Interfacial Phenomena in Apolar Media*, (Eicke and Parfitt, Eds.), Marcel and Dekker Inc., 1985, Chapter 8.

34 B.J. Briscoe, P.S. Thomas, and D.R. Williams, Microscopic origins of the interface friction of organic films: the potential of vibrational spectroscopy, *Wear*, **53** (1), 263, 1992.

35 B.J. Briscoe, Interface wall friction in powder flows, *Proc. NATO Conf. Physics of Granular Media*, (Bideau and Dodds, Eds.), 1992.

36 F.P. Bowden and D. Tabor, *Friction and Lubrication of Solids*, Oxford University Press, **I**, 1950 and **II**, 1964.

37 D.F. Moore, *Principles and Applications of Tribology*, Pergamon Press, Oxford, 1975.

38 A. Cameron and W.J.S. Grew, *Proc. R. Soc. London*, **A186**, 179, 1972.

39 D. Dowson, *History of Tribology*, Longmans, New York, 1979.

40 B.J. Briscoe, M.J. Adams, and T.K. Wee, The differential friction effect of keratin fibres, *J. Phys. D. Appl. Phys.*, **23**, 406, 1990.

41 G.A.D. Briggs and G.J. Briscoe, How rubber grips and slips: Schallamach waves and the friction of elastomers, *Phil. Mag*, **A38** (4), 387, 1978.

42 B.J. Briscoe, *Friction of organic polymers, fundamentals of friction: Macroscopic and microscopic process*, (I.L. Singer and H.M. Pollock Eds.), Kluwer Academic Publishers, 1992, pp. 167–182.

4

Adhesive Wear

WILLIAM A. GLAESER

Contents

4.1 Introduction

It has been shown in Chapter 2 that real surfaces in contact can develop strong adhesive bonds. Generally, the extent of native oxide films and adsorbed species, as well as the plastic yield properties of the solids, dictate the possibility of adhesive bonding. Since real materials are bumpy on a microscopic scale, real contact occurs at only a few high points. Concentration of the load in a few small contact areas results in high stresses in these regions. These asperity contacts are where adhesion is likely. To move the two surfaces, then, requires breaking the adhesive bonds or shearing subsurface material. The results can lead to transfer of material from one surface to the other, resulting in adhesive wear.

Fortunately, nature makes adhesion of contacting surfaces difficult. Atomically clean metal surfaces have unsatisfied bonds that would immediately bond strongly to another "clean" surface. The bonds would be similar to a grain boundary. In our environment, however, surfaces have low energy states because of reacted and adsorbed species that satisfy the loose bonds. Our skin is a low energy surface that does not stick to anything. In fact, we have fine grooves or fingerprints on our fingers so that we can hold onto things without their slipping out of our grasp.

Figure 4.1 Microtopography of a worn steel surface (SEM micrograph)

When we put materials in high vacuum environment and drive off adsorbed gas molecules by heating, we can produce adhesion when we bring the materials into contact. Before the space age, there were speculations that bearings operating in outer space would weld together. Consequently, in anticipation of space probe operations, sliding contact experiments were conducted in high vacuum chambers. To those researchers' surprise, producing contact adhesion was more difficult than anticipated. Without instrumentation to measure the surface state at that time, the conditions leading to adhesion or cold welding could not be determined.

Bearings and gears do not involve just bringing surfaces into contact. Sliding contact involves plowing of asperities through mating surface material. Inspection of worn surfaces shows long parallel scratches and "ironing out" of ridges as shown in Figure 4.1. Microplowing creates a state of stress different from normal contact. Near-surface shear produces plastic flow parallel to the surface. Soft surface films are scraped off and brittle oxides are broken up and plowed under. This can bring clean regions into contact and introduces the possibility of strong adhesions. This process is diagrammed in Figure 4.2.

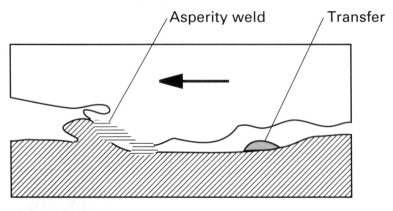

Figure 4.2 Asperity plowing

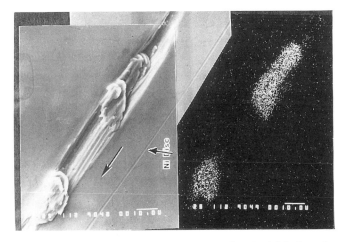

Figure 4.3   In-SEM wear experiment showing the initiation of a transfer event

In any contact system there is a critical stress above which the probability of adhesion increases significantly. Therefore, adhesion and transfer will occur initially in a random way. In addition, the clean surface patches are competed for by gases in the environment. Depending on the rapidity with which stable adsorption occurs, adhesion may be inhibited by *self healing* films—depending on the system involved.

A sequence of asperity contact, adhesion, and transfer is shown in Figure 4.3. This sequence is from a series of in-SEM wear experiments conducted to observe the initiation of wear.[1] The sequence follows a plowing mark that appears on the moving surface, its adhesion, and shearing off. The plowing mark appears in the upper right hand corner; the arrow parallel to the adhesion mark shows the direction of travel of the contacting asperity. This experiment involved a nickel disk and an iron pin. A dot pattern EDX scan for iron is shown on the right, indicating the areas of iron transfer to the disk. Subsequent passes over the transfer lump tended to smear out the lump and enlarge its area. Note that there is another plow mark in the field of view (indicated by arrow) which did not develop a transfer. The original local transfers roughen both surfaces and induce more interaction, creating a cascade of events that produces a discontinuous transfer layer.

Chemisorption of gas molecules from the atmosphere is known to profoundly affect dry adhesive wear. The chemical affinity of a given metal species for oxygen or nitrogen determines the effectiveness of presence of the gas. Mishina[2] has developed Table 4.1 for several classes of metals, showing their affinities for oxygen and nitrogen.

The transition metals, iron, titanium, nickel, and cobalt, have nonpaired electrons in the d orbital. This leads to high chemisorption activity. As the $3d$ orbital

Pure metal	Chemisorption activity
Iron	Chemisorbs O_2 and N_2
Titanium	Same
Nickel	Chemisorbs O_2 but not N_2
Cobalt	Same
Zinc	Chemisorbs O_2 weakly, does not chemisorb N_2
Gold	Does not chemisorb either gas

Table 4.1   Chemisorption activity of metals

fills from titanium, with 2 electrons, to nickel, with 8 electrons, the affinities for the two gases changes.

Mishina performed wear experiments in atmospheres containing various partial pressures of the two gases, oxygen and nitrogen. The results were quite interesting. When strong chemisorption occurred, maximum wear did not occur under high vacuum conditions, as might be expected, but rather at a pressure of about 10^{-1} Pa. Increasing the atmospheric pressure above that value brought a rapid decrease in the wear rate. For gold, the wear rate was not influenced by atmospheric pressure. It is suggested by Mishina that at high vacuum conditions adhesion occurs, but debris sticks to the surfaces and is not ejected, so that the apparent wear is small. As the amount of gas in the atmosphere is increased, the wear particles adsorb gas and do not stick to the surfaces or each other as readily, and therefore are ejected from the surfaces as relatively fine debris. Further increase in gas pressure produces an adsorbed gas layer sufficient to inhibit any adhesion and the wear decreases significantly. The author emphasizes the point that wear does not decrease because of the growth of an oxide film, since X-ray diffraction analysis of the debris showed no evidence of oxide. One might argue about what constitutes "oxide" and what constitutes chemisorbed oxygen. The surface chemistry could better be defined with Auger or XPS analysis.

4.2 Surface Analysis

Adhesive wear involves asperity adhesion and transfer. Mechanical aspects of this process involving contact stress and plastic deformation are fairly well understood. The surface chemistry of the process, however, has only begun to be fathomed. For instance, the interface between the transfer material and the substrate is of critical importance. Is it a "grain boundary?" Is there a thin oxide film or a monolayer of

ADHESIVE WEAR Chapter 4

adsorbed gas? Or is the interface a mixture of patches of film and exposed substrate? The kinds of bonds involved are of interest also.

It is important, therefore, not only to model the state of stress and the near-surface microstructure resulting from sliding contact but also to define the chemistry of the surfaces. This means "taking snapshots" not only of the surface chemistry during contact and adhesion, but also of the changes that occur in the surface chemistry.

Most of the research to date has dealt with the use of surface analytical tools to explore the surface state after the fact. This research has provided new and valuable insights into the mechanisms of adhesive wear. Some examples will be given.

4.3 Auger Analysis of Worn Surfaces after "Unlubricated Wear"

Thomas and Glaeser[3] considered the possibility that in unlubricated wear not all asperity contacts experienced adhesion and transfer. Friction, then, might be related to the combined effects of shearing of occasional adhesive contacts and the plowing of hard asperities through softer surface material. SEM and EDX analysis might not show evidence of adhesion and transfer because the amount of transfer is too small to be detected by these means. This type of transfer might even occur under boundary lubrication conditions.

Using a steel ring rubbing against a copper–aluminum alloy block, wear scars were produced under dry sliding conditions and under lubricated conditions. The wear scars developed on the copper aluminum blocks were analyzed in a Perkin-Elmer model 595 scanning Auger spectrometer with a spatial resolution of less than 100 nm. SEM mode surface microscopy did not reveal unambiguous evidence of iron transfer. An SEM micrograph of a typical area in the copper aluminum block wear scar is shown in Figure 4.4. Accompanying Auger spectra keyed on iron are shown in Figure 4.4. Three spots, a, b, and c noted on the photomicrograph are analyzed in the scans. The high-resolution Auger used here showed that the areas of iron transfer on the surface were extremely small.

The lubricated wear scars contained dark patches showing high concentrations of carbon and oxygen. A micrograph in Figure 4.5 shows a typical surface area. Iron was found within these patches (a), but was not found in areas free of dark patches (b). Sputter cleaning the surface removed most of the visible "organic" patches, but some iron was still detected. These findings correlated with previous analyses of wear debris—that in lubricated wear, wear debris seems to be concentrated in an organometallic material. It was also noted that considerable aluminum was present in the dark patches. Comparing the Al/Cu ratio in the patches and on dry surface showed an increase in aluminum concentration of about 5 times. The analysis suggested that aluminum was being preferentially leached out of the alloy during the lubricated wear process.

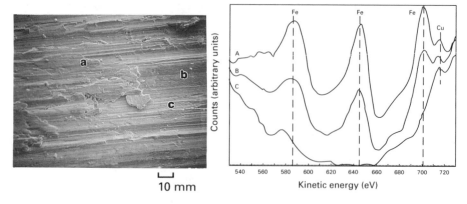

Figure 4.4 Ion Auger scans taken at the three locations shown in the micrograph. The spectra shown were taken in the N(E)E mode of data acquisition. The scale factor for each curve is the same, and they are displaced vertically for clarity

These analyses showed that, indeed, transfer can occur between contacting solids and not be detected by conventional methods (microscopy). They also demonstrated that although wear may occur during lubrication, it can occur without adhesive transfer. The debris can be captured in an organic deposit, but there is no evidence of direct transfer of metal from one surface to the other at asperity contacts. An unexpected result of the analyses was the discovery that an alloy constituent can be leached out of the surface by the lubrication process.

Singer and coworkers at the US Naval Research Laboratory in Washington, DC have studied the wear of TiN-coated tool steel.[4] Using a ball-on-flat tester, the TiN coating was rubbed and the ball surface analyzed for transfer. The ball materials were steel and sapphire. Auger analysis of the steel ball surfaces revealed metallic iron, iron oxide, and titanium oxide. The sapphire ball was coated with titanium

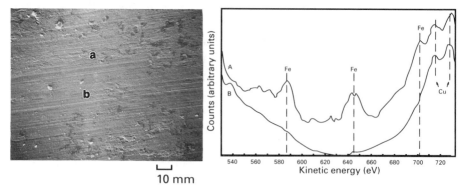

Figure 4.5 Iron Auger spectra of the two areas shown in the micrograph after 40 nm ion sputter etch

ADHESIVE WEAR   Chapter 4

oxide only. The transfer material was stripped off the ball surfaces and examined by transmission electron microscopy. The material was crystalline, and was identified by electron diffraction as a $FeTiO_3/\alpha\ FeO_3$ mixture and TiO_2. This work lead to the conclusion that oxidation of the TiN coating surface produces a low shear strength film which transfers to the mating surface. During later stages of the wear process, back transfer from one surface to the other occurs. These experiments illustrate how surface oxide can control adhesive wear under mild sliding conditions.

In experiments reported by Rigney and Vankstesan,[5] sliding steel surfaces in air produced debris mostly composed of oxide. In addition, no transfer was detected. In these experiments one of the steel parts of the sliding pair was activated and transfer was determined by measuring the spectra of gamma rays emitted and looking for the Co-56 decays using a Ge–Li detector.

4.4 *In Situ* Systems

Systems have been developed to enable observation of wear processes in microscopic detail and to analyze surface chemical changes as they occur. Observations of microscopic details of wear in the SEM have been carried out by a number of researchers.[6–11] Some of these experiments are described in Chapter 5. *In situ* systems generally involve high vacuum environment, and therefore will not reflect the powerful effects of air environment on the wear process. SEMs have been modified so that the filament electron source can be kept at high vacuum while the specimen is essentially at ambient conditions. This has allowed observation of sliding contact in the presence of a film of liquid lubricant.[12]

In spite of the vacuum environment, there are a number of advantages for *in situ* systems. It might be argued that the same thing could be done with light microscopy in the air. However, the superior depth of focus of SEMs makes it possible to see microtopographical features on the wear scar that are not seen in light microscopy. Further, with SEMs equipped with WDS or EDX elemental analysis systems, transfer material can be identified as it is laid down, and the physical changes in transfer patches with continuous sliding can be followed. Carbides can be located and identified in the surface so that their particular role in the wear process can be observed.

In keeping with the insertion of wear test apparatus into surface measuring instruments, there have been attempts to do the same with scanning Auger systems. Buckley and associates[13] experimented with dry sliding pin-on-disk systems to detect transfer of various metals. Transfer of iron, nickel, and cobalt to tungsten was reported, while no transfer of iron or nickel to molybdenum was detected.

Glaeser[14] has investigated the transfer of lead to steel from leaded bronze alloy, using an in-Auger wear apparatus. A photograph of the apparatus is shown in Figure 4.6.

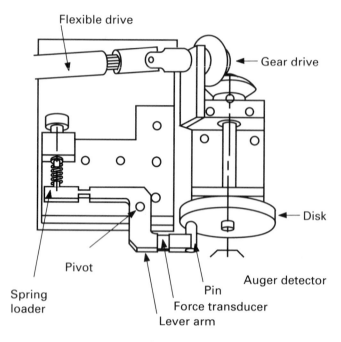

Flexible drive

Gear drive

Disk

Auger detector

Pin

Force transducer

Lever arm

Pivot

Spring
loader

Figure 4.6   In-Auger wear apparatus

The apparatus is designed in the pin-on-disk configuration. The wear apparatus is mounted on a standard stage and vacuum flange for an Auger spectrometer. As shown in Figure 4.6, the disk is mounted with its axis horizontal on a shaft supported by a pair of rolling contact bearings. The shaft is driven through an angle gear drive with a flexible driver which penetrates the vacuum flange. The prime mover is a variable speed dc electric motor mounted on the exterior of the flange. A

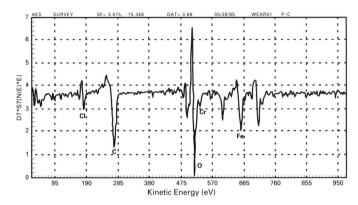

Figure 4.7 **Auger spectrum of stainless steel wear surface taken before sliding contact took place**

pin with a rounded nose is mounted on a pivoting load arm. A spring loading system is used to load the pin against the disk. (The dark rectangle near the pin is a reflection of the load arm in the polished surface of the disk.)

The load arm is strain gauged to measure normal load and frictional force. A spring is used to load the arm. The wear system is mounted on a spectrometer specimen holder with remote control x, y, and z positioning mechanisms. Thus the surface of the disk can be positioned so that the Auger detector can pick up the Auger electrons from the wear track. Remote position controls allow adjustment of the relative position of the pin on the disk surface so that multiple wear tracks can be developed once the assembly is mounted in the spectrometer.

The wear experiments were performed in a Perkin-Elmer Model 560 scanning Auger spectrometer having a spatial resolution of about 1 μm. The sliding speed can vary from 0.04 m/min. During the leaded bronze experiments, the load was 4.4 N. The environment was high vacuum, 10^{-9} torr.

Lead transfer experiments have demonstrated the importance of the native oxide structure on the surface receiving the transferred material. A leaded bronze pin was rubbed against a stainless steel (304 stainless) rotating disk in the Auger spectrometer. Prior to initiating sliding contact, the disk surface was analyzed, and the spectrum shown in Figure 4.7 was attained. The spectrum showed oxygen, carbon, chlorine, chromium, and iron peaks. The carbon and chlorine represent ubiquitous contamination usually found on metal surfaces, even after careful cleaning. The oxygen signature is typical of oxide, and represents the chromium oxide on the stainless surface. Lead transfer was detected after a number of disk rotations. This is shown in Figure 4.8. An SEM micrograph of the lead transfer zone is shown in Figure 4.9. The amount of lead is minute. Sputter profiling the lead transfer indicated a film thickness of about 1.5–2.0 nm. The width of the lead track averaged 0.02 mm. Lead globules in the bronze microstructure were about

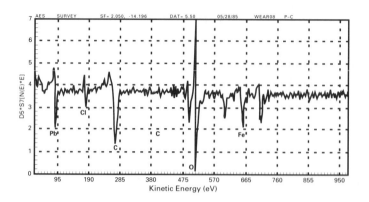

Figure 4.8   Auger spectrum of wear scar region containing lead transfer

0.3 mm in diameter. It was presumed that the lead transfer came from one lead globule. The sputter profile is shown in Figure 4.10. It can be seen that when the lead concentration drops, the oxygen level increases rapidly. It was assumed from these results that lead had transferred to the oxide film on the stainless steel surface.

Additional observations made in these *in situ* experiments were that no transfer occurred until the carbon and chlorine levels in the wear track were considerably reduced. This showed that the normal hydrocarbon contamination on the metal surface is sufficient to inhibit adhesion. Further, when the stainless steel surface was sputter cleaned until the carbon, chlorine, and oxygen levels were considerably reduced, transfer of bronze occurred. The effect of the oxide layer on control of the

Figure 4.9   SEM photomicrograph of lead transfer area on stainless steel wear surface. Lead is revealed as a white streak

ADHESIVE WEAR   Chapter 4

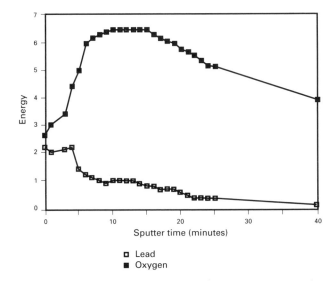

Figure 4.10   Ion sputter profile of lead transfer zone

metal species transferred was demonstrated. The reason why lead transferred and bronze did not when oxide was present is still to be determined.

Further *in situ* experiments have been carried out with a mild steel disk instead of a stainless steel one. The results were different from the stainless steel results. Lead did not transfer from the leaded bronze, while oxide remained on the disk surface. Furthermore, when transfer occurred, both bronze and lead were found. No purely lead transfer regions could be developed.

The surface chemistry of the wear scar on the disk was followed as a function of the number of revolutions of the disk. Figure 4.11 shows the surface chemistry of the as-received disk, showing a strong peak for carbon as well as oxide and iron peaks. The hydrocarbon layer is thick enough to partially hide the iron oxide layer

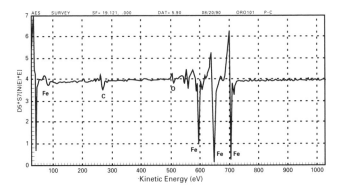

Figure 4.11   Auger spectrum for as-received steel disk

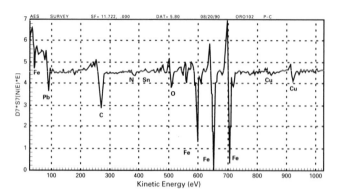

Figure 4.12 Auger spectrum of disk wear scar after 1 revolution

under it. After just 1 revolution of the disk, a significant reduction in carbon was observed. This is shown in the spectrum in Figure 4.12. Note the increase in the strength of the oxygen signal. After 15 revolutions the spectrum shown in Figure 4.13 shows further reduction of the carbon signal, but no indication of lead or copper. Runs were then made in a spectrometer with higher spacial resolution. A plot of atom ratios of O/Fe, C/Fe, and Cu/Fe as a function of the number of revolutions is shown in Figure 4.14. The reduction in carbon can be seen as quite rapid. After 40 revolutions, when the carbon and oxygen levels were substantially reduced, significant quantities of copper were detected as the number of rotations increased to 60. Thus bronze transfers to the steel surface after hydrocarbons and oxygen are substantially removed by rubbing contact.

4.5 Conclusions

The previous examples of the use of surface analytical techniques for the study of wear mechanisms have shown that, for adhesive wear, we have only told a small part

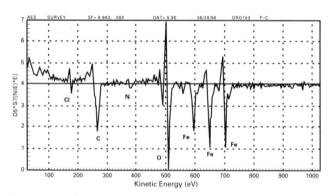

Figure 4.13 Auger spectrum of wear scar after 15 revolutions

ADHESIVE WEAR Chapter 4

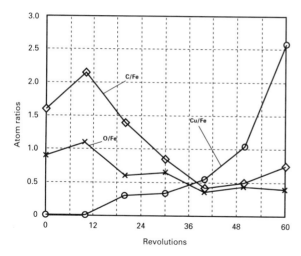

Figure 4.14 Composition changes in the steel surface as a function of disk revolutions

of the story. These works have shown small aspects of the process, which added to the already established understanding still does not provide a comprehensive theory of adhesive wear.

Just changing materials and operating conditions produces new surface chemical phenomena that apparently are part of the wear mechanism. We know that adsorbed gases can attenuate the adhesion process. (Even hydrogen has been found to change the friction of carbon-graphite.) Native oxides or oxides grown during the rubbing process can combine between contacting surfaces and produce a product of changed composition. Some oxide growth can be controlled by grain boundary diffusion. Hence, in the case of pearlite colonies in steel, oxide will grow faster on the pearlite than on the ferrite phase since the pearlite has a much finer grain structure. Oxide growth is slower on pearlite lamellae normal to the surface as compared to lamellae orientated parallel to the surface. The wear process can disrupt this structural effect.[15] If we accept the role that surface chemistry plays in modifying the adhesion process, we must also consider the reaction kinetics involved. How fast must an oxide layer grow to keep up with the rubbing away of it? What happens to the rubbed-off oxide—does it leave the contact zone as debris, is it transferred back and forth between the surfaces, or is it just pushed aside to, perhaps by chance, rejoin the sliding system?

Ferrante and Bozzolo in Chapter 2 point out that adsorbed films don't always reduce adhesive forces. Oxygen adsorption on single crystal sapphire increased the interfacial shear strength, while chlorine reduced it. Hartweck and Grabke[16] have demonstrated in some closely controlled gas adsorbtion studies that a partial monolayer of oxygen or nitrogen will increase the adhesive strength of an iron–iron contact in an ultrahigh vacuum chamber. Sputter-cleaned iron surfaces are brought into contact in the vacuum system, and the force required to separate them is measured. A 30% monolayer of oxygen produces a fourfold increase in average adhesive bond strength as compared with the as-sputtered surfaces. Further increase in the percent monolayer causes a rapid decrease in adhesive strength. The authors propose that the Fe–O–Fe bonding is stronger than an Fe–Fe bond formed by pressing two surfaces together.

Engineering surfaces are contaminated with hydrocarbons and water vapor. These layers sit on top of the oxide layers. *In situ* Auger wear experiments have demonstrated that even in high vacuum conditions these contaminants can effectively shield the contacts from adhesion—at least until they are scraped off.

The contact state of stress is another factor which contributes to the system. If contact stress is large, subsurface shearing takes place. Surface films can be broken up, or plowed under. Localized adhesions are probabalistic, depending on the chance encounters between clean areas. Large localized strain associated with wear will also modify the surface chemistry. It has been shown that alloy constituents can be leached out of the surface during wear, leaving a segregated material with different chemical potential than the original surface.

Transfer during adhesive wear can produce a transfer layer of different structure and composition than the bulk. Rigney[17] has shown that a transfer layer composed of a mechanically mixed material with 10–20 nm particles or subgrains can develop during unlubricated wear of copper. The chemical behavior of this new material can be expected to differ from the copper substrate.

In addition to the importance of the mechanical properties of the surface and near-surface material, the effects of large strain and frictional heating on the microstructure of the surface material must be taken into account. In fact, the hardness of the contacting surfaces is a powerful determinator for wear resistance. A hard oxide layer on a soft metal surface (oxide on aluminum, for instance) may be ineffective because the soft substrate will deform, allowing the oxide to break up like ice on a pond.

High strain in the near-surface region can cause phase transformations to occur in some materials. For instance, 304 stainless steel has an austenitic structure which is unstable and will transform to hard martensite during heavy deformation. On the other hand, 316 stainless steel has a very stable austenite structure which does not transform during the wear process.[17]

References

1 W.A. Glaeser, Wear experiments in the scanning electron microscope, *Wear*, **78**, 371–386, 1981.

2 H. Mishina, Atmospheric characteristics in friction and wear of metals, *Wear*, **152**, 99–110, 1992.

3   M.T. Thomas and W.A. Glaeser, Surface chemistry of wear scars, *J. Vac. Sci. Technol.*, **A2** (2), 1097–1101, 1984.

4   I.L. Singer, S. Fayeulle, and P.D. Ehni, Friction and wear behavior of TiN in air: The chemistry of transfer for films and debris formation, *Proceedings of International Wear of Materials Conference*, ASME, 1991, 229–242.

5   S. Venkastesan and D.A. Rigney, Sliding friction and wear of plain carbon steels in air and vacuum, to be published in *Wear*.

6   D.H. Buckley, *Surface Effects in Adhesion, Friction, Wear and Lubrication*, Elsevier Tribology Series, **5**, 54–59, 1981.

7   K. Hokkirigawa and K. Kato, Theoretical estimation of abrasive wear resistance based on microscopic wear mechanism, *Wear of Materials*, ASME, 1–8, 1989.

8   S.V. Prasad and T.H. Kosel, *In Situ* SEM scratch tests on white cast irons with round quartz abrasive, *Wear of Materials*, 121–129, 1983.

9   Y. Tsuya, K. Saito, R. Takagi, and J. Akaoka, *In Situ* observation of wear process in a scanning electron microscope, *Wear of Materials*, 1979, 57–71, 1979.

10 W.A. Glaeser, *op. cit.*

11 S.V. Pepper and D.H. Buckley, NASA TDN-6497, 1971.

12 RPI SEM experiments with liquid lubricants.

13 S.V. Pepper and D.H. Buckley, NASA TND-6497, 1971.

14 W.A. Glaeser, Transfer of lead from leaded bronze during sliding contact, *Wear of Materials*, 255–280, 1989.

15 Venkastesan, *op. cit.*

16 W. Hartweck and H.J. Grabke, Effect of adsorbed atoms on the adhesion of surfaces, *Surface Science*, **89**, 174–181, 1979.

17 D.A. Rigney, L.H. Chen, M.G.S. Naylor, and A.R. Rosenfield, *Wear*, **100**, 195–219, 1984.

18 D.A. Rigney, Some thoughts on sliding wear, *Wear*, **152**, 187–192, 1992. To be published in the Kragelskii Memorial Volume on *Friction and Wear*, USSR Academy of Sciences.

5

Abrasive Wear

KOJI KATO

Contents

5.1 Abrasive Asperities and Grooves

In standard sliding tests, parallel grooves are generally formed on wear surfaces after sliding, even in the case of contact between smooth surfaces of the same material. This means that the microstructure of a surface is not uniform, and the mating microspots on surfaces have different hardnesses.

Figure 5.1 shows the hardness distribution on a wear surface of 0.45% carbon steel sliding against the same material in a pin-on-ring test.[1]

The initial surface hardness (before sliding begins) varies by about 200 kgf/mm^2 to 260 kgf/mm^2 depending on the place of indentation. But after sliding about 30 m the hardness variation increases to between about 300 kgf/mm^2 and 700 kgf/mm^2.

The difference in the degree of microscopic work hardening on wear surface generates a wide distribution of microhardness.

Figure 5.2 shows a cross section of two mating surfaces (a pin and a ring) which have the hardness distribution[1] of Figure 5.1. It is obvious from the figure that the two surface profiles conform to each other.

This type of sliding situation can be called abrasive, and the peak surfaces are harder than the valley surfaces. When surfaces of dissimilar materials are in sliding

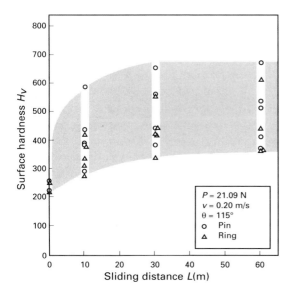

Figure 5.1 Microhardness on wear surfaces of 0.45% carbon steel

contact, the microscopic difference in hardness on wear surface should be large to produce true abrasive wear.

Thus, the microscopic contact situation must be as schematically shown in Figure 5.3, where one groove is formed by one asperity which has an *attack angle* of θ. If the attack angle is changed, the wear rate drastically changes, as shown in Figure 5.4, where the sudden increase of the wear rate of a ring is observed as a function of the attack angle.[1]

In this way, the microscopic hardness distribution on mating surfaces forms abrasive asperities and grooves, and the attack angle of asperity is confirmed to be important in abrasive wear.

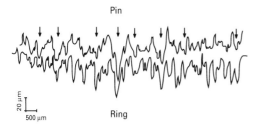

Figure 5.2 Cross sections of the mating wear surfaces of 0.45% carbon steel after pin-on-ring sliding

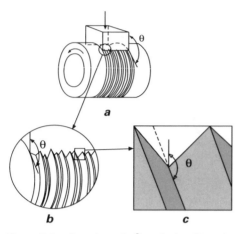

Figure 5.3   Attack angle θ at the leading edge of a pin specimen

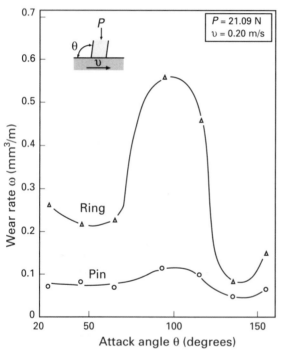

Figure 5.4   Effect of attack angle on wear rate of 0.45% carbon steel in unlubricated sliding

5.2 Yield Criterion of an Abrasive Asperity

An asperity of hardness H_1 can form a groove on a flat surface of hardness H_2 only when the asperity does not yield under the combined forces of normal load and tangential friction.

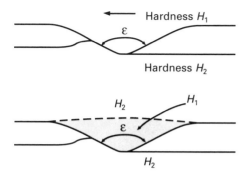

Figure 5.5 Hardness ratio r and asperity tip angle at abrasive contact point

Theoretical analysis[2] with slip line field theory on the yielding conditions of a two-dimensional asperity in abrasive sliding shows that asperity yield is determined by a combination of the hardness ratio $r \, (= H_2/H_1)$, the tip angle ε, and the dimensionless interfacial shear strength f (the interface shearing strength/bulk shearing strength of a flat), which are shown in Figure 5.5. In the case $f = 0$, the critical tip angle ε_c for the asperity yield is given by the following equations as a function of hardness ratio r:

$$\varepsilon_c = (2r + \pi - 2)/2 - r, \quad r \geq 4/(4+\pi)$$
$$r = (2 - 2\cos\varepsilon_c)/2 + \varepsilon_c, \quad r \leq 4/(4+\pi) \tag{5.1}$$

Figure 5.6a shows the theoretical relationship between ε_c and r calculated using Equations 5.1, where a groove is formed by an abrasive asperity only when the tip angle ε is larger than ε_c. When ε is smaller than ε_c the asperity yields, and effective grooving is not performed.

Figure 5.6b shows the relationships between ε_c and r when f varies from zero to unity. It is clear from the figure that an asperity yields more easily under smaller value of f and larger hardness ratios.

Abrasive Wear Mode Diagram

Microscopically, when an asperity is strong enough to cause grooving of the mating surface, the abrasive action removes surface material in various ways as abrasive wear, and the groove or scratch remains as a surface feature in the microtopography.

Thus, the microwear mechanisms in abrasive wear should provide a fundamental understanding of the sliding wear process for metals.

In order to study these micromechanisms as they occur, an *in situ* wear system was developed for the scanning electron microscope. In some of the first *in situ* SEM wear studies, various types of wear particles and their generating processes

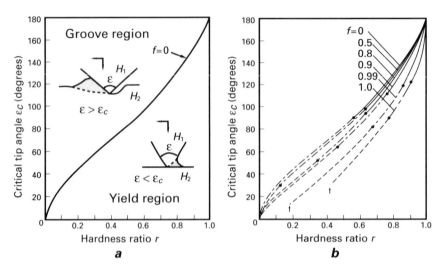

Figure 5.6 Relationship between the critical tip angle and the hardness ratio *r* where shearing strength changes from 0 to 1.0

were carefully observed by several investigators.[3–10] The effect of repeated sliding on surface grooving and the initiation of wear were also observed.[11, 12] Subsequent *in situ* SEM wear experiments included the measurement of friction, and plastic theory was introduced for dynamic analysis of wear modes.[13–15] The wear mode diagram was introduced as the result of these analyses and *in situ* SEM wear experiments.

Classification of microscopic wear modes and the introduction of the wear mode diagram stimulated the analysis of the wear mode transition process in repeated sliding. A video system was connected to the scanning electron microscope for this purpose, and the process of wear mode transition in repeated sliding was analyzed with many successive observations.[16] Thus the *in situ* wear study evolved from the single pass sliding condition to the multipass sliding condition, which gave more practical information about the wear mechanisms.

Observations of wear processes under lubrication have also been carried out by Holzhauer and Calabrese. A standard scanning electron microscope was modified with a valve–gauge system, and good-resolution images of liquid-lubricated surfaces were produced routinely.[17] These observations show that *in situ* SEM studies of wear under lubrication with grease or oil will provide a new aspect for understanding the lubricated wear mechanism.

Figure 5.7 shows a drawing of the friction apparatus, in which the friction pair consists of an upper pin specimen and a lower flat specimen.[18] The apparatus is placed in the scanning electron microscope with the dead weight load in place and the pin in contact with the disk. With the chamber closed and under vacuum, the

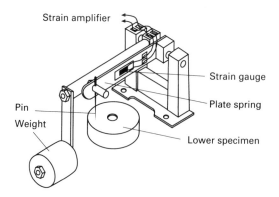

Strain amplifier

Strain gauge

Plate spring

Pin

Weight

Lower specimen

Figure 5.7   *In situ* **SEM wear apparatus**

disk is rotated and the wear scar is observed at high magnification as it forms. The friction force is measured by the strain gauge assembly attached to the plate spring. Two different steel pin specimens were used, having point radii of 27 μm and 62 μm, respectively. Brass (60Cu–40Zn), 0.45C steel, and austenitic stainless steel were used for the lower disk specimen. The surfaces were finished by buff polishing followed by cleaning with acetone in an ultrasonic cleaner. Six dead weight loads were used, ranging from 0.21 N to 1.0 N. Figure 5.8 shows representative abrasive wear models observed in the SEM in the sliding process of a steel pin against a brass disk.[19] In Figure 5.8a a long ribbon-like wear particle is generated, and this wear mode is called the *cutting mode*. It causes a large wear rate and is observed under larger loads and smaller hardness ratios. A deep and long groove is formed distinctly in this case.

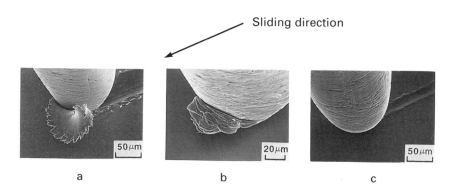

Sliding direction

50 μm   20 μm   50 μm

a   b   c

Figure 5.8   **Three abrasive wear modes observed by SEM: a) cutting mode, steel pin on brass plate; b) wedge forming mode, steel pin on stainless steel plate; c) ploughing mode, steel pin on brass plate**

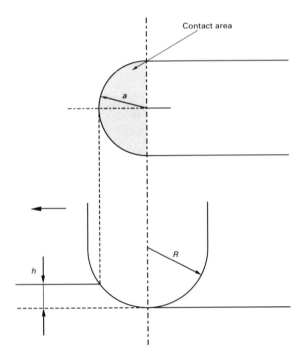

Figure 5.9   Model of contact between a spherical pin and a flat surface:
_h_, depth of groove; _a_, radius of contact area, _R_, radius of a pin

In Figure 5.8b, a wedge is formed ahead of the pin and sticks to it. Once this wedge is formed, sliding takes place at the interface between the wedge bottom and the flat surface in the following process. This wear mode is called the _wedge forming_ mode, which is observed when adhesion between a pin and a flat is strong. The wear rate under this wear mode is much smaller than that of cutting mode, and the groove surface is generally rough, with transverse cracks.

In Figure 5.8c, a wear particle is not clearly observed at the interface and only a shallow groove is formed plastically on a flat surface. This wear mode is called a _ploughing mode_, which is observed when adhesion at the interface is weak and the penetration of an abrasive asperity is relatively small. Although a shallow groove is formed as a result of plastic deformation, real volume removal from the surface does not take place in a single pass of sliding.

In order to describe the severity of contact of an abrasive which causes different modes of wear, the model in Figure 5.9 is introduced for the contact between a spherical pin and a flat specimen. The degree of penetration D_p is defined by the following equation:

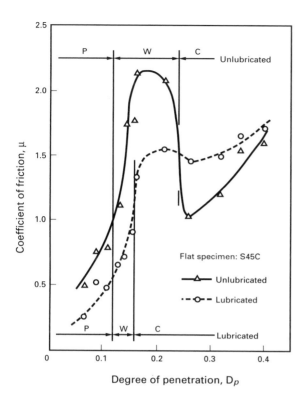

Figure 5.10 Relationships between friction coefficient μ and degree of penetration D_p for steel pin sliding against stainless steel disk, lubricated and unlubricated: P, ploughing mode; W, wedge forming mode; C, cutting mode

$$D_p = \frac{h}{a} = R\sqrt{\frac{\pi H_v}{2W}} - \sqrt{(\frac{\pi H_v}{2W}) R^2 - 1} \qquad (5.2)$$

where h is the depth of penetration, a is the radius of contact area, R is the radius of the pin tip, W is the normal load, and H_v is the hardness of a flat specimen.

The friction coefficient μ changes depending on the value of D_p, which is shown in Figure 5.10 together with the wear mode. Although a change in the friction coefficient is complicated by the change in D_p, it can be explained theoretically if D_p is translated into an attack angle θ in a two-dimensional abrasive model. This angle is shown in Figures 5.3 and 5.4, and is equal to $(\pi-\varepsilon)/2$ in Figure 5.5.

The two-dimensional theoretical analysis with slip line field theory gives the following relationships[20] for the cutting mode:

$$\mu = \tan \left(\theta - \frac{\pi}{4} + \frac{1}{2}\cos^{-1}f\right) \tag{5.3}$$

the wedge forming mode:

$$\mu = \frac{\left[1 - 2\sin\gamma + \sqrt{(1 - f^2)}\right]\sin\theta + f\cos\theta}{\left[1 - 2\sin\gamma + \sqrt{(1 - f^2)}\right]\cos\theta - f\sin\theta} \tag{5.4}$$

and the ploughing mode:

$$\mu = \frac{A\sin\theta + \cos(\cos^{-1}f - \theta)}{A\cos\theta + \sin(\cos^{-1}f - \theta)}$$

$$A = \left(1 + \frac{\pi}{2} + \cos^{-1}f - 2\theta - 2\sin^{-1}\right)\left(\frac{\sin\theta}{\sqrt{1 - f}}\right) \tag{5.5}$$

where γ is the angle determined by the stress discontinuity.

Equations 5.3–5 explain very well the observed relationship between μ and D_p in Figure 5.10 when D_p and θ are related geometrically by the following equation:

$$D_p = 0.8\tan\frac{\theta}{2} \tag{5.6}$$

In this way, the theoretical relationships between D_p and f for the wear mode transitions are obtained by Equations 5.5–6 and are shown in Figure 5.11 by solid lines, together with *in situ* SEM experimental observations.[21]

The solid lines distinctly divide the possible wear mode regions of cutting, wedge forming, and ploughing. It is shown in the figure that the wedge forming mode appears at larger values of f and that the transition between the cutting mode and the ploughing mode is directly affected by D_p, especially at smaller values of f.

It is shown later that Figure 5.11 is very useful for the further development of a wear theory; it is called the abrasive wear mode diagram.

5.3 Degree of Wear at One Abrasive Groove

In order to know the amount of wear volume produced in one pass of an abrasive asperity, the measurement of the cross sectional profile of a groove is useful. As shown in Figure 5.12, the apparent area of a groove does not totally give the wear volume. Such wear grooves are schematically shown in Figure 5.13, where $\Delta A'$ means the cross sectional area of a valley and $\Delta A''$ is the cross sectional area of side ridges.

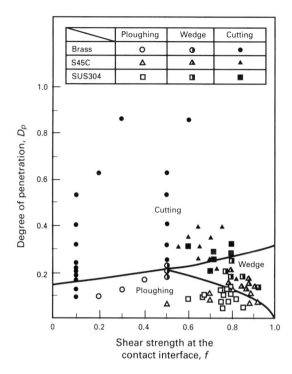

Figure 5.11　Abrasive wear mode diagram: wear modes as a function
of degree of penetration D_p and shear strength at the
contact interface f

The difference between $\Delta A'$ and $\Delta A''$ corresponds to the wear volume caused by
the unit sliding distance, and β, defined by the following equation, gives the degree
of wear at one groove:

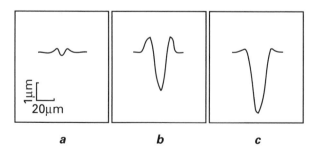

Figure 5.12　Cross sectional profiles of grooves; a) ploughing mode, b) wedge forming mode,
and c) cutting mode

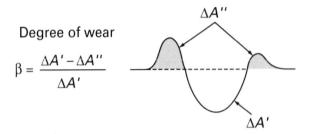

Degree of wear

$$\beta = \frac{\Delta A' - \Delta A''}{\Delta A'}$$

Figure 5.13 Definition of degree of wear β at one groove

$$\beta = \frac{(\Delta A' - \Delta A'')}{\Delta A'} \tag{5.7}$$

When there exists no wear but only plastic deformation, β becomes zero, and when the volume of a groove is totally removed and $\Delta A''$ is zero, β becomes unity.

Figure 5.14 shows the change of β caused by the change of D_p in the case of a steel pin sliding against a steel disk of different hardness. In the figure, the wear modes of cutting, wedge forming, and ploughing are shown by different symbols.

It is quickly understood from these results that β varies between 0.8 and 1.0 in the cutting mode, between 0.1 and 0.8 in the wedge forming mode, and is almost zero in the ploughing mode.

It is also shown in Figure 5.14 that the cutting mode of large β appears at smaller D_p values as the hardness of material increases, and that the wedge forming mode appears for a wide range of D_p values as the hardness of material decreases.

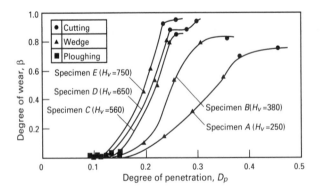

Figure 5.14 Degree of wear β as a function of degree of penetration D_p observed with heat treated steel (H_v is the hardness)

5.4 Macroscopic Wear in Multiple Abrasive Sliding Contacts

In the macroscopic contact between two flat surfaces, there are multiple asperities in contact at the interface. Although they have different shapes and different contact pressures, the contact pressure is generally supposed to be constant and is taken to be equal to the hardness H_v.

If there are N abrasive asperities in contact and each abrasive contact has the projected area ΔA, then ΔA is given by the following equation:

$$\Delta A = \frac{W}{NH_v} \tag{5.8}$$

In order to translate the projected area ΔA into the cross sectional groove area $\Delta A'$, the shape factor C is introduced; ΔA and $\Delta A'$ are related as follows:

$$\Delta A' = C\Delta A \tag{5.9}$$

Since the wear volume ΔV at one abrasive groove after the sliding distance L is defined as follows,

$$\Delta V = \Delta A' \beta L \tag{5.10}$$

ΔV is described as follows by substituting Equation 5.8 into Equation 5.10,

$$\Delta V = C\beta \frac{WL}{NH_v} \tag{5.11}$$

On the other hand, the shapes of abrasive asperities vary over a wide range, and the contact configuration occurs as a result. If the shape of an asperity is modeled by a regular pyramid, the attack angle θ formed by the abrasive front surface of the pyramid and the abraded flat surface can be a good representative value for determining the wear mode.

It was already shown in the abrasive wear mode diagram of Figure 5.11 that three wear modes appear depending on the values of θ and the shearing strength f at the interface. For example, the critical attack angles θ_{c1} and θ_{c2} for wear mode transitions are determined by introducing the proper value of f_0 as shown in Figure 5.15.

The distribution of θ in practical abrasives is shown in Figure 5.16 as an example. By substituting the values of θ_{c1} and θ_{c2} obtained in Figure 5.15 into Figure 5.16, the fraction of abrasive contacts in the cutting mode is given by α_c, and the fraction in the wedge forming mode is given by α_w. The remaining fraction corresponds to the ploughing mode.

Figure 5.17 shows the relationships between α_c, α_w, and f obtained in this way. By taking these values of α_c and α_w into consideration, the real contributive num-

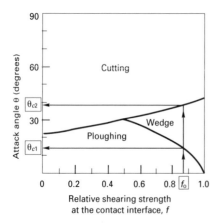

Figure 5.15 Determination process of the critical attack angle for the wear mode transition on abrasive wear mode diagram

bers of abrasive contacts in the number of N are given by $\alpha_w N$ for the wedge forming mode and $\alpha_c N$ for the cutting mode.

Therefore, the total wear volume V at N abrasive contacts is introduced from Equation 5.9 as follows:

$$V = N\Delta V = (C_w \beta_w \alpha_w + C_c \beta_c \alpha_c) \cdot \frac{WL}{H_v} \tag{5.12}$$

The contribution of the ploughing mode to the wear volume is neglected since its value of β is very small, as is seen in Figure 5.14.

Microscopic experimental analysis gives the following empirical values for the coefficients in Equation 5.12 for standard alloys and pure metals:

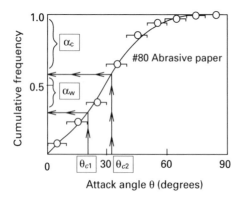

Figure 5.16 Distribution of the attack angle of alumina abrasive grains

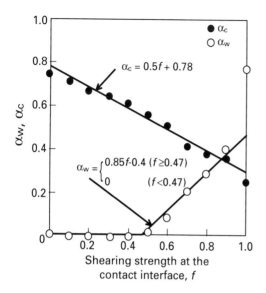

Figure 5.17 Relationship between α_w or α_c and f

$$C_w = 0.10$$
$$C_c = 0.13$$

$$\beta_w = 0.20 \sim 0.80 \quad (\text{mean value: } 0.50)$$
$$\beta_{wc} = 0.80 \sim 0.95 \quad (\text{mean value: } 0.88)$$

$$\alpha_w = \begin{bmatrix} 0 & (f < 0.47) \\ 0.85f - 0.40 & (f \geq 0.47) \end{bmatrix}$$
$$\alpha_c = -0.5f + 0.78 \tag{5.13}$$

Figure 5.18 shows the experimental and theoretical relationships between wear resistance and hardness for various metals, where wear resistance is defined[22] as WL/V. The solid line shows the theoretical values calculated using Equation 5.12 and the experimental values mentioned above. Data for standard alloys and pure metals were observed in tests with abrasive papers.

It is confirmed from the results that Equation 5.12 shows good agreement with the macroscopic wear resistance of alloys and pure metals. In the case of heat-treated steels, the cutting mode becomes predominant, and both β_c and α_c depend on the hardness.

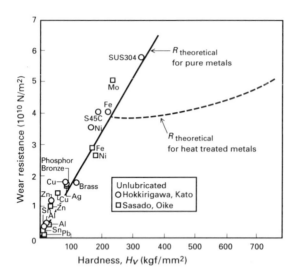

Figure 5.18 Relationship between abrasive wear resistance and hardness of worn material observed experimentally and estimated theoretically

Figure 5.19 shows that the degree of wear β_c in the cutting mode increases with hardness, and Figure 5.20 shows that the fraction of abrasive contacts α_c increases with hardness.

In the calculation of theoretical (broken) lines in Figure 5.18 for heat-treated steels, the values of β_c and α_c in Figures 5.19 and 5.20 were used, and the term corresponding to wedge forming in Equation 5.12 was neglected since its contribution

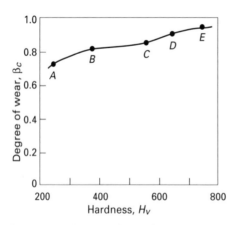

Figure 5.19 Degree of wear for cutting mode as a function of hardness, observed with heat-treated steel

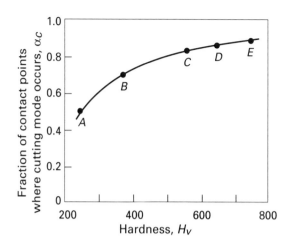

Figure 5.20 Fraction of abrasive contacts where cutting mode occurs, as a function of hardness, observed with heat-treated steel

is relatively small. The shape factor C_c was determined so as to have good fitting with the macroscopic wear resistance of iron and S45C in Figure 5.18.

The theoretical estimations shown by the solid line and the broken line in Figure 5.18 agree well with other well-known observations for abrasive wear.[23] Therefore, the *in situ* SEM microscopic analysis, with the model of an abrasive asperity and the concept of the abrasive wear mode diagram, seems to give a good explanation of the abrasive wear mechanism. The degree of wear at one abrasive groove is also shown to be an important parameter for understanding the wear mechanism.

References

1 Y.C.Chiou and K. Kato, Wear mode of micro-cutting in dry sliding friction between steel pairs (Part 1), *J.JSLE*, **32**, 41–48, 1987.

2 T. Kayaba, K. Kato, and K. Hokkirigawa, Theoretical analysis of the plastic yielding of a hard asperity sliding on a soft flat surface, *Wear*, **87**, 151–161, 1983.

3 T.R. Bates, K.C. Ludema, and W.C. Brainard, A rheological mechanism of penetrative wear, *Wear*, **30**, 309, 1974.

4 W.A. Brainard and D.H. Buckley, Dynamic SEM Wear Studies of Tungsten Carbide Cermets, *ASLE Trans.*, **19**, 309, 1976.

5 T. Kayaba and K. Kato, The analysis of adhesive wear mechanisms by successive observation of the wear process in SEM, ASME, *Proc. Int. Conf. on Wear of Materials*, 1979, 45.

6 Y. Tsuya, K. Saito, R. Takagi and J. Akaoka, *In situ* observation of wear processes in a scanning electron microscope, ASME, *Proc. Int. Conf. on Wear of Materials*, 1979, 57.

7 T. Kayaba, K. Kato, and Y. Naggasawa, Abrasive wear in stick-slip motion, ASME, *Proc. Int. Conf. on Wear of Materials*, 1981, 439.

8 S.J. Calabrese, F.F. Ling, and S.F. Murray, Dynamic wear tests in the SEM, *ASLE Trans.*, **26**, 455, 1983.

9 L. Ahman and A. Oberg, Mechanism of micro-abrasion *in situ* studies in SEM, ASME, *Proc. Int. Conf. on Wear of Materials*, 1983, 112.

10 S.C. Lim and J.H. Brunton, A dynamic wear rig for the scanning electron microscope, *Wear*, **101**, 81, 1985.

11 W.A. Glaeser, Wear Experiments in the scanning electron microscope, *Wear*, **73**, 371, 1981.

12 S.V. Prasad and T.H. Kosel, *In situ* SEM scratch tests on white cast iron with round quartz abrasive, ASME, *Proc. Int. Conf. on Wear of Materials*, 1983, 121.

13 T. Kayaba, K. Kato, and K. Hokkirigawa, Theoretical analysis of plastic yielding of a hard asperity sliding on a soft flat surface, *Wear*, **87**, 151, 1983.

14 T. Kayaba, K. Hokkirigawa, and K. Kato, Experimental analysis of the yield criterion for a hard asperity sliding on a soft flat surface, *Wear*, **96**, 255, 1984.

15 T. Kayaba, K. Hokkirigawa, and K. Kato, Analysis of the observed wear process in a scanning electron microscope, *Wear*, **110**, 419, 1986.

16 H. Kitsunai, K. Kato, K. Hokkirigawa and H. Inoue, The transition between microscopic wear modes during repeated sliding friction by SEM-tribosystem, *Wear*, **135**, 1990.

17 W. Holzhauer and S.J. Calabrese, Modification of SEM for *in situ* liquid-lubricated sliding studies, *ASLE Trans.*, **30**, 302, 1986.

18 T. Kayaba, K. Hokkirigawa, and K. Kato, Analysis of the abrasive wear mechanism by successive observations of wear processes in a scanning electron microscope, *Wear*, **110**, 419, 1986.

19 K. Hokkirigawa and K. Kato, An experimental and theoretical investigation of ploughing, cutting and wedge formation during abrasive wear, *Tribology International*, **21** (1), 51–57, 1988.

20 J.M. Challen and P.L.B. Oxley, An explanation of the different regimes of friction and wear using asperity deformation models, *Wear*, **53**, 229–243, 1979.

21 K. Hokkirigawa and K. Kato, Theoretical estimation of abrasive wear resistance based on microscopic wear mechanism, *Wear of Materials*, 1–8, 1989.

22 M.M. Khruschov, Resistance of metals to wear by abrasion—as related to hardness, *Proc. Int. Conf. on Lub. and Wear*, Inst. Mech. Engr. London, 1957, 655–659.

6

Boundary Lubrication

K.C. LUDEMA

Contents

6.1 Introduction

Two clean solids will adhere or stick together simply by bringing their surface atoms toward each other within a distance of one or two atomic spacings. This is the normal state of matter, and it does not require heating, rubbing, or any action other than bringing the surfaces close together. Intervening substances of the right kind prevent close proximity, which limits the strength of adhesion. For example, one atomic layer of adsorbed nitrogen on each surface reduces the adhesion between two rough steel surfaces to less than 5% of that for truly clean surfaces. In normal atmosphere, steel surfaces are covered with several nanometers of oxide and adsorbed gas, which reduces the potential for adhesion still more, to almost negligible values. But,

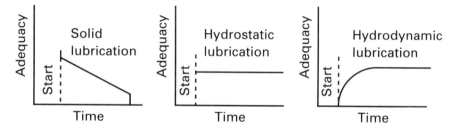

Figure 6.1 Adequacy of lubricant over time of operation

adsorbed gas layers can be "pushed" aside by squeezing and rubbing to increase adhesion.

Any substance placed between two sliding solids which lowers their sliding resistance or adhesion could be called a lubricant—so broad is the definition of lubrication. Solids, liquids, and gases are all effective lubricants, each in particular applications. Some lubricant–metal pairs produce chemical effects that increase lubricant effectiveness, and others do not. The latter lubricate by mechanical means only.

6.2 Mechanical Effects in Lubrication

Apart from the materials used to lubricate, there are several strategies used to lubricate surfaces. Some of these may be described using the sketches in Figure 6.1. Solid lubricants (or soft coatings) are applied on hard substrates before the start of sliding. After the start the lubricant film wears away until there is too little to protect the sliding surfaces from damage. In most cases, it is not practicable or practical to replenish a solid lubricant, and thus the sliding system has limited life. This limited life may, however, be adequate for special tasks such as are required in some space vehicles.

A second type of system, hydrostatic lubrication, pumps lubricant (e.g., gas, liquid, or grease) into the contact area at sufficient pressure to separate the sliding surfaces. The sliding pair can function immediately and can continue to function without failure. In the third case in Figure 6.1, the system begins immersed in lubricant, but it must start-up in order to develop a lubricant film and separate the sliding pair. One direct analogy is water skiing. This is called hydrodynamic lubrication.

6.3 Adequacy of Hydrodynamic Fluid Films

The mechanical adequacy of a fluid film may be expressed in terms of Λ, which is defined as h_0/σ. The quantity h_0 is the film thickness, and the roughness, σ is most reasonably taken as $(\sigma_1^2 + \sigma_2^2)^{1/2}$. These σ values are the roughnesses of contacting surfaces 1 and 2, which are statistical expressions of asperity height. When $\Lambda > 3$ there is virtually no contact between asperities. At values of Λ nearer 1 there is some contact and wear.

The value of h_0 can be calculated for the hydrodynamic case by several equations found in the literature. (Listing of these equations is beyond the scope of this article, and may be found in textbooks on lubrication.) One early equation was developed for the contact condition between two cylinders of radius R and length L rolling against each other with an applied load of W. Each cylinder rotates at a different speed with an average surface velocity of U, immersed in a lubricant of viscosity η:[1]

$$\frac{W}{L} = 2.45\frac{UR\eta}{h_0} \tag{6.1}$$

This equation shows one important point which applies to machinery that must be stopped and later restarted: at $U = 0$, $h_0 \Rightarrow 0$, which condition produces very high friction and wear.

Conversely, fluid film bearings accommodate shock loads and short-term overload fairly well. These cases may be treated as squeeze film problems. If a sliding pair is operating with sufficient fluid film and a pulse load is applied, the fluid film will not immediately diminish in thickness. Rather, time is required to squeeze the film to some inadequate dimension. Thus, very short pulses may not cause bearing damage at all. The equation governing this behavior is of the type of Stefan's equation, which gives the rate at which a circular disk of radius a and weight W falls through a liquid of viscosity η in a direction perpendicular to the plane of the disk; h is the spacing between the disk and the bottom of the container holding the liquid:

$$\frac{dh}{dt} = \frac{2h^3 W}{3\eta\pi a^4} \tag{6.2}$$

One way of characterizing adequacy of lubrication is as variations in the coefficient of friction, in the form of the Streibeck curve. The shape is shown generally in Figure 6.2, where the abscissa is written in terms of Z, the viscosity of the lubricant; N, the rotational speed of a shaft in a bearing; and p, the nominal contact pressure between the shaft and bearing. Traditional regimes of lubrication are labeled.

Apart from the high friction at start-up, sliding systems can be designed such that asperities never touch during sliding. The result is usually a heavy and costly machine. The competitive drive in making consumer products led to the discovery that many practical sliding systems function well though values of Λ might sometime decrease to less than 1. This was found to be due to the formation of protective

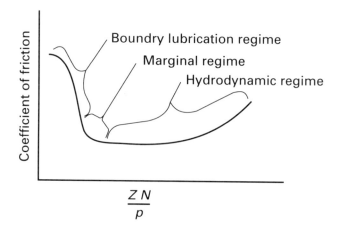

Figure 6.2   Streibeck curve, showing three regimes of lubrication over a range of lubricant viscosity *Z*, sliding speed *N*, and contact pressure *p*, between a shaft and bearing

films on surfaces by reaction between the sliding surfaces and active constituents in the oil. This phenomenon was first connected with the use of minimally refined oil.[2]

With the discovery of ways to assure such chemical effects in lubricants, manufacturers gained confidence in designing products using smaller bearings than usual, which could be overloaded or poorly maintained now and again, or started and stopped often, without significant compromise of useful life.

6.4 Chemical Effects in Liquid Lubrication— Boundary Lubrication

Studies of the chemical effects in lubrication began before the year 1900. Early papers are largely anecdotal in nature, describing how test results depended on the source of lubricants, the history of use of the lubricant, the metals in the bearings, and on how the bearings were made, to name a few variables. In the 1920s Hardy[3] demonstrated several chemical effects and suggested that this mode of lubrication be referred to as *boundary lubrication*. At the same time, other authors reported that there must be a level of hydrodynamic performance on the asperity level that was not readily calculated. Some authors have referred to this mixed state as *quasi-hydrodynamic lubrication.*

Research in the 1930s showed that there are several separable effects in marginal lubrication. In order to separate some of these effects, virtually all of the early research on boundary lubrication was done[4] "by sliding surfaces over one another at extremely low speeds and very high contact pressures so that the incidence of

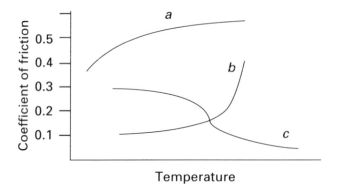

Figure 6.3 Behavior of three types of lubricant over a range of temperatures, at low sliding speed and high contact pressures: a) an unreactive lubricant/-metal pair, in which lubricant viscosity decreases and μ rises as temperature increases; b) a fatty acid, or the metal soap of a fatty acid, which melts at a particular temperature and becomes ineffective as a lubricant; and c) a liquid lubricant containing a reactive constituent that forms low shear strength compounds on the sliding surface which endure to high temperature. These reactive constituents are referred to as *EP* additives, denoting their effectiveness under conditions of *extreme pressure* of contact

hydrodynamic or elastohydrodynamic lubrication [was] reduced to a minimum." The test specimen shapes in that day were the pin-on-disk and the four-ball geometries. Sliding of inert liquid–metal pairs shows that friction rises as temperature increases, as shown[5] by curve *a* in Figure 6.3. This is attributed to a reduction of viscosity in the lubricant as temperature rises.

The effect shown in curve *b* is simply due to the difference between the effectiveness of a *grease* (fatty acid) and a liquid (melted fatty acid). The behavior shown by curve *c* is due to chemical reaction of a lubricant with a *reactive* solid. The effects shown in curve *c* would not occur when using chemically *inert* solids as sliding surfaces, such as most polymers or many ceramic materials.

6.5 Wear and Failure

When $\Lambda > 3$ there is no asperity contact and yet wear can occur when abrasive materials enter the system. For values of Λ nearer 1 there will be some asperity contact and some wear even without abrasives. Asperity contact can be verified by measuring electrical resistance between conducting sliding pairs. When Λ is 1 or slightly less, in a nonreactive lubricant–slider system, there is much more wear. With polymers and with soft metals, particularly single-phase metals, surface roughness increases

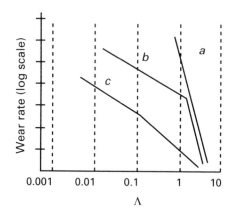

Figure 6.4 Wear rate of steel in laboratory tests using three lubricants,[6] at ≈ 20°C: a) vacuum pump oil, sliding in an atmosphere of N_2; b) laboratory grade mineral oil, in air atmosphere; and c) commercial gasoline engine oil, in air atmosphere

quickly, leading to very high friction and wear. This is referred to as scuffing, scoring, or other terms. For values of $\Lambda \Rightarrow 0.8$ one can expect severe wear and early failure with nonreactive lubricant–slider systems.

Some coatings or films form on surfaces by reaction with the general environment to form oxides or other compounds. These films are protective; that is, they prevent seizure at low sliding speeds. Modern lubricants contain additives specifically to react with steel (mostly) to form films for higher speed use.

The compounds that are formed by chemical reaction reduce the coefficient of friction and allow operation at values of Λ well below 1, perhaps to as low as 0.01 for short periods of time. These films include ions of the substrate metal. As these compounds are rubbed off or worn away some metal ions depart, which constitutes wearing away of metal. The rate of chemical reactivity therefore becomes important. If the desired chemical reaction rate is slow, the supply of protective film may not keep pace with the rate of loss. On the other hand, the chemical reaction rate could be so high that substrate material is corroded away at a greater rate than would ordinarily be lost during sliding. Further, the products of corrosion are usually poor lubricants and cause high wear rates. Such chemical balance is the goal in the formulation of lubricants for metal cutting, engines, gears, etc.

The comparison of the effectiveness of several lubricants is shown schematically in Figure 6.4. These are the results of laboratory tests described in Figures 6.7–9 below. Among other things, Figure 6.4 shows that the wear rate in boundary lubrication is not some constant fraction of the wear rate in dry sliding, as is often suggested in the literature.

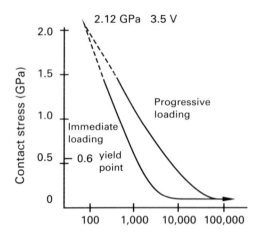

Figure 6.5 Durability of steel sliding surfaces, lubricated with substances containing reactive constituents, comparing systems where full loads were applied immediately and progressively[6]

6.6 Surface Protection When $\Lambda < 1$—Break-In

Most engineers know that new surfaces cannot be immediately operated at low values of Λ, whereas they function well after some *break-in*. Makers of mechanical components often break-in machines by operating them gently at first, or in some cases with a special oil, sometimes containing a fine abrasive. Each strategy has its own purpose. An abrasive compound in oil simply laps the sliding surfaces into conformity. Some break-in oils contain a more chemically active additive than normal, to accelerate the formation of films in regions of high contact stress. However, if these oils are left in the system they might cause excessive corrosion.

The effectiveness of break-in may be seen in some laboratory tests. Tests were done to determine how long a lubricated surface would survive when loads were applied by two methods, namely, by immediate loading and by progressive loading, both at constant speed.[6] The results are shown in Figure 6.5. At a contact pressure of 0.2 GPa the survival time with progressive loading is twenty times that with immediate loading. At 0.8 GPa there is a tenfold difference, and at 1.5 GPa the difference is only four times. Both results converge to a single point at a contact pressure in excess of 2.1 GPa or $\approx 3.5\,Y$, where Y is the tensile yield strength of the metal. The benefits of a break-in procedure appear to diminish as loads increase. There is also a contact pressure below which the value of Λ is great enough to totally avoid wear.

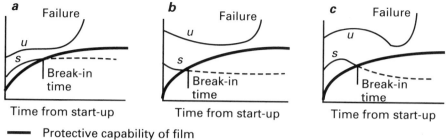

Figure 6.6 Comparison of protective capability of films (which improves with time) versus the requirements of the sliding surfaces, each graph showing surfaces that can (*s*) and cannot (*u*) become satisfactorily protected by the developing surface film: a) surfaces that become worse with sliding; b) surfaces that become better with sliding, for a time; and c) more variable sliding surfaces

6.7 Dynamics of Break-In

A crucial juncture in new, that is, previously unused lubricated machinery is the first start-up. Protective films do indeed build up, but this requires some time of sliding. New surfaces are at risk until some protection develops. Most surfaces are not deliberately coated before start-up, but to prevent failure, compounds are sometimes applied to new surfaces, which are referred to as break-in coatings. These are usually zinc or iron phosphates. Their exact role has never been determined. Some authors suggest that these compounds have rough surfaces, which "trap" and hold lubricant until other protective films can develop. Others suggest that these compounds simply have lower shear strength than does a substrate metal, and function as solid lubricants.

The dynamics of break-in have not been well studied. Break-in is often thought to be a smoothing effect, which would be expected to make the existing lubricant film more effective. The problem with this view is that some surfaces become rougher during break-in than they were originally, and they function very well also after break-in. We may suggest, however, that some balance of events occurs as illustrated in Figure 6.6.

In Figure 6.6 are plotted two competing events for three cases, the protective capability (i.e., film thickness) of chemically formed films versus the requirement of the sliding surfaces. Both change with time of sliding. A surface as manufactured has some initial requirement in lubrication, according to its surface condition, hardness, and doubtless other properties. In the relatively unprotected start-up of sliding, surfaces change in some way (e.g., roughness), which usually increases the requirements for lubrication, as shown in Figure 6.6a. If the protective capability of films overtakes the changing surface requirements, the surfaces will survive. Otherwise the surface will fail. Figures 6.6b and 6.6c show two other cases of surface change during break-in.

6.8 Research in Boundary Lubrication

There are many papers on boundary lubrication, and yet there are few basic principles by which lubricants may be selected and bearings designed. Surely the major reason is that boundary lubrication is very complicated, and depends on many variables. These include chemical variables, materials variables, manufacturing variables, and variables in the operation of bearings. Thus among the great number of published papers, most cover a very small part of the entire field. Some authors work with full-scale machinery, but are constrained to study them under either very limited and controlled conditions, or under such general conditions that the data can only be analyzed by statistical methods. Laboratory methods are not entirely satisfying because of the tenuous connection between the behavior of laboratory devices and the behavior of practical machinery.

Perhaps the most confusion in the literature is in determining when boundary lubrication exists. It would seem most logical that the *prima facie* evidence for boundary lubrication would be the existence of an adequate boundary film. The problem is that adequate films are usually not visible, nor measurable by standard laboratory methods. Thus authors define boundary lubrication in other ways. For example, a machine is sometimes said to be operating in the state of boundary lubrication if its lubricant contains active additives. Others follow the changes of friction as shaft speed decreases or as bearing load is increased: if the coefficient of friction passes through some minimum as shown in Figure 6.2, then the system must have entered the boundary lubrication regime. Still others will declare that if the coefficient of friction is greater than about 0.1 the system must be operating in the boundary regime.

6.9 Laboratory Research

The experiments discussed here used a cylinder-on-flat geometry in step–load tests. A sketch of the specimen shapes is shown in Figure 6.7. The research device correlates fairly well with cams–followers in gasoline engines. Tests were run in a step–load mode which approximates endurance tests. Figures 6.8 and 6.9 show how the two type of tests may be run. Figure 6.8 is a sketch of endurance (time to failure) versus applied load in a lubricated sliding test. The longest time to failure, at low load may be 30 days. At high load the time to failure may be as short as 30 seconds.

A much shorter test procedure is the step–load test, conducted as shown in Figure 6.9. The adequacy of lubrication is measured by the load at which surface

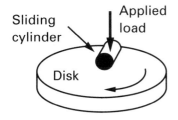

Figure 6.7   Cylinder-on-flat test geometry

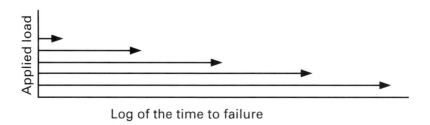

Figure 6.8   Endurance tests

failure occurs. There is a good correlation between time to failure in the endurance test and the load of failure in the step–load test.

Step–load tests were used to learn the influence of several variables on lubricant durability. These variables include surface roughness, sliding speed, metal hardness and microstructure, break-in procedures, and temperature. Results using steel and various lubricants will be reported for insight into the influence of system variables. Further, the chemical compositions of the films from these tests were determined and the mechanical durability of the boundary films was measured.

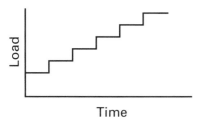

Figure 6.9   Step–load test

6.10 Composition of Films

Films form on surfaces of many lubricated mechanical components, but the most studied components have been automotive engines, transmissions and other components. Most attention has been paid to the widely-used additive zinc-dialkyl-dithio-phosphate (ZDP). Chemically oriented papers usually describe the films to consist mostly of oxides and friction polymer. Zinc phosphide, zinc phosphate, zinc sulfate, zinc sulfide, or iron phosphate have been specifically ruled out.[7] Most papers attempt to make a connection between the chemical composition within the lubricant itself and its function as a lubricant. Engine oils contain mostly the original hydrocarbons, paraffinic or napthenic, plus the thermal oxidation products including alcohols ($R-$ $-OH$), ketones ($R-C[=O]-R$), acids ($R-C[=O]-OH$), and esters ($R-C[=)]-O-R$). In addition to ZDP, there are other additives for acid buffering, foam suppression, dirt suspension, etc. The function of an oil as a scuff prevention agent is clearly only a part of the concern of chemists.

The films formed on lubricated steel surfaces include a flake form of Fe_3O_4, upon which is an organo-iron compound (OIC).[8] (For a description of the measurement of film thickness and estimation of composition see the Appendix on Ellipsometry.) The oxide flakes are a few nanometers thick and less than a micrometer across. The coefficient of friction of a dry steel surface covered with this oxide is about 0.12, whereas the clean dry steel has a coefficient of friction of about 0.25.

Relatively little OIC forms in laboratory grade mineral oil used as a lubricant, and it is ketone- and acid-based. With the addition of ZDP, less oxide forms but the OIC is ester-based and becomes thicker. It contains up to 15% total P+Zn+S. For the same conditions of boundary lubrication, the mineral oil allows between 24 and 80 times more wear than does formulated engine oil. This is probably because the OIC with P+Zn+S effectively covers the relatively hard Fe_3O_4 flakes.

Figures 6.10a–c show that the films require time to develop, and the rate of growth is influenced by the hardness of the steels. These figures suggest two distinct layers in the films: actually, the lowest part of the film is exclusively oxide, the upper part is exclusively organo-iron and there is a gradation of composition between.

The films can sustain limited contact severities, seen as load in Figure 6.10. When the load reaches about 550 N the films begin to diminish in thickness and would fail if the load were not relieved. After the load is relieved, the films build up again. This sequence shows that break-in and scuffing are not described by instantaneous values of surface temperature or Λ, but rather by the time required to build up or diminish the films.

Several other phenomena were observed. One is that the thickness of oxide film is considerably thicker in oil containing 1% ZDP than for either 0.5% or 2% ZDP. ZDP does not increase the rate of film formation, but it does produce a film that will withstand about 3 times the load sustained by the films formed from mineral oil. Further, the rate of formation of films varies considerably in commercial

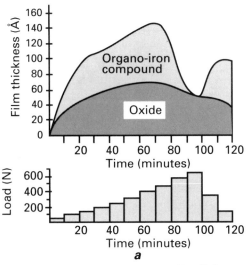

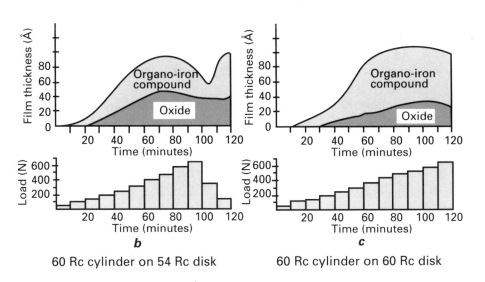

Figure 6.10 **Idealized two-film layers on steel disks of three hardnesses, with mineral oil lubricant, 0.06 m/s sliding speed, step–loading as shown in the bar graphs[8]**

lubricants over a range of temperatures, reaching a maximum at temperatures in the range of 200°C to 300+°C. Another finding was that monolithic, i.e., furnace grown Fe_3O_4 is not effective in reducing friction. Finally, Fe_2O_3 forms in water- and air-saturated oil, whereas Fe_3O_4 forms even in deaerated water, and ZDP suppresses the formation of Fe_2O_3. These are all strong indicators of the complexities of the chemistry of film formation.

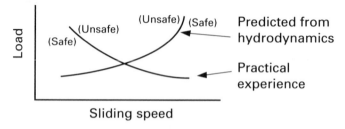

Figure 6.11 Comparison of the expected variation in load-carrying capacity of boundary lubricated films compared with practical experience

6.11 Further Mechanical Effects of the Boundary Lubricant Layer

In hydrodynamic theory, the higher the sliding speed the thicker the fluid film should be. However, it is more often found that the load-carrying capacity of lubricated systems is lower at high speed than at low speed. These conditions are shown in Figure 6.11. Several reasons have been given for this, including the dynamic effects of bearing misalignment. The more probable reason is that at high sliding speeds the asperities are heated adiabatically, which makes them soft and readily deformed. The surfaces roughen quickly at high sliding speeds and fail because the state of lubrication, which is already marginal, is even less effective against the rougher surface.

Surface roughness has an effect on break-in in addition to that expressed in the value of Λ. With soft steel it was found in laboratory experiments that under the conditions of the test, proper breaking-in by sliding requires a specific initial surface roughness of about 0.1 μm Ra. Smoother and rougher surfaces failed quickly, as shown[9] in Figure 6.12. It appears that the optimum surface roughness is one in which the asperities plastically deform at a rate that is too slow for fast progression

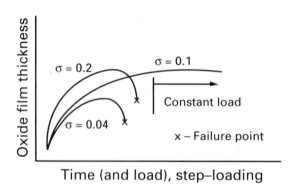

Figure 6.12 The film thickness change during a step load test, as influenced by the surface roughness, Ra[9]

BOUNDARY LUBRICATION   Chapter 6

to low-cycle fatigue failure of surface metal, but at a rate sufficient to accelerate the formation of oxides. The surface roughness of the original smooth and rough surfaces stayed the same throughout most of the tests. The surfaces with intermediate surface roughness became smoother.

6.12 Surface Analysis of Boundary Lubricated Metals

There is no standard appearance of boundary lubricated surfaces. Rather, the appearance depends strongly on the type of metal used and the state of progression toward failure. A change in surface roughness during sliding is quite surely a manifestation of a state of boundary lubrication, but some boundary lubricated surfaces do not change roughness. However, the appearances of surfaces change.

Poor bearing materials, such as brass, soft stainless steel, and aluminum alloys will usually have the appearance of local galling or adhesion. Good bearing materials will usually have the appearance of local smearing, perhaps of a soft second phase. These features are often of the scale of size of the metal grains.

Zones of galling and zones of smearing are often "lined up" to produce long marks. These marks are often interpreted as abrasion marks, leading to the conclusion that the surface may have been worn in an abrasive mode. There is a significant difference between abrasion scratches and the marks from boundary lubrication. In abrasion, the furrows often have materials built up alongside the furrows and will often have relatively steep walls. In functioning, i.e., not failing, boundary lubrication, the marks or lines are of a smaller scale than are abrasive scratches and there is usually no material built up along the marks. It is necessary to observe the surfaces at magnifications between 50× and 250× to see the distinction. During such observation, it will be seen that counter surfaces of different hardness or microstructure have different appearances. The marks on each surface will differ according to the nature of the contacting material but also according to the nature of the films that form on each surface. The most useful approach to surface analysis is to obtain considerable experience in observing surfaces, over a wide range of magnification and with binocular microscopes, polarizing microscopes, and electron microscopes.

References

1 H.M. Martin, Lubrication of gear teeth, *Engineering* (London), 102 and 199, 1916.
2 A.W. Burwell, *Oiliness*, Alox Chemical Company, Niagara Falls, 1935.
3 W.B. Hardy, *Collected Works*, Cambridge Press, 1936.
4 F.P. Bowden and D. Tabor, *Friction*, Anchor/Doubleday, 1973, 120.

5 F.P. Bowden and D. Tabor, *The Friction and Lubrication of Solids*, Oxford University Press, 1950.

6 Y.Z. Lee and K.C Ludema, The shared load wear model in lubricated sliding, *Wear*, **138**, 13–22, 1990.

7 A. Beerbower, Boundary lubrication, *ASLE Trans.*, **14**, 90–104, 1971.

8 B. Cavdar and K.C. Ludema, Dynamics of dual film formation in boundary lubrication of steels, part I: functional nature and mechanical properties, *Wear*, **148**, 305–327, 1991; part II: chemical analysis, *Wear*, **148**, 329–346, 1991; part III: real-time monitoring with ellipsometry, *Wear*, **148**, 347–361, 1991.

9 S.C. Kang and K.C. Ludema, The breaking-in of lubricated surfaces, *Wear*, **108**, 375–384, 1986.

Appendix: Ellipsometry and Its Use in Measuring Film Thickness

Effective breaking-in of lubricated steel surfaces has been found to be due primarily to the rate of growth of protective films of oxide and compounds derived from the lubricant. The protection afforded by the films is strongly dependent on lubricant chemistry, steel composition, original surface roughness, and the load–speed sequence or history in the early stages of sliding. Given the great number of variables involved it is not possible to follow more than a few of the chemical changes on surfaces using electron microscopes and other analysis instruments at the end of the experiments. A method was needed to monitor surfaces during experiments and in air. Ellipsometry was used for real-time monitoring, and the detailed analysis was done by electron-, ion-, and X-ray-based instruments at various points to calibrate the results from the ellipsometer.

A complete description of ellipsometry may be found in various books, and the particular ellipsometer used in the work mentioned in this article is described in Cavdar and Ludema.[8] Fundamentally, ellipsometry makes use of various states of polarized light. The effect of a solid in changing the state of polarized light is now described.

Polarized light is most conveniently described in the terms of the wave nature of light. Plane-polarized light is simply represented as a sine wave on a flat surface, as shown in Figure 6.13. The "end view" of the wave may be sketched as a pair of arrows. Light is directed toward a surface at some chosen angle relative to a reflecting surface, called the *angle of incidence*, as shown in Figure 6.14. If the surface is no rougher than about λ/10, the light will reflect with little scatter, which is referred to as *specular reflection*. Linearly polarized light may be directed upon a surface at any angle of rotation or *azimuth*, relative to the plane of incidence, as shown in

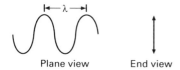

Plane view   End view

Figure 6.13 Wave of radiation, in plane view and end view

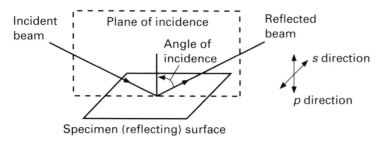

Incident beam

Plane of incidence

Angle of incidence

Reflected beam

s direction

p direction

Specimen (reflecting) surface

Figure 6.14 Light beam incident upon and reflecting from a surface, showing the plane of incidence and the directions of the s and p components of light

Figure 6.15. The incident light can be represented as having separated into two components, the s component and the p component, as shown in Figures 6.14 and 6.15.

Each component is treated differently by the reflecting surface: each component changes in both phase and intensity at the point of reflection, as shown in Figure 6.16. The sketch shows an incident plane-polarized beam at an azimuth of 45°, which can be thought of as separating into two equal components. Each component is changed in both intensity and phase at the point of reflection. The

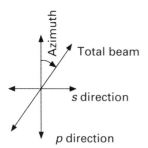

Azimuth

Total beam

s direction

p direction

Figure 6.15 One possible orientation of a plane or linearly polarized beam relative to the plane of incidence (which contains the p component)

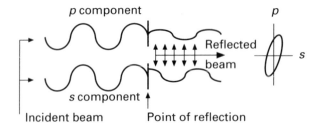

Figure 6.16 Two separate influences of a reflecting surface on the *s* and *p* components of incident plane polarized light. Incident beam is plane-polarized at an azimuth of 45°, making the *s* and *p* components equal. The reflected beam is seen to be elliptically polarized. The ellipse can be constructed by plotting the intensities of the *p* and *s* components at various positions along the waves

reflected components then recombine to form a beam that is polarized elliptically. The elliptical shape may be seen by plotting the *s* and *p* components, both at the same time or point in the reflected waves, over a wavelength.

Formally, the changes in intensity of each of the *s* and *p* beams, and the phase shift δ of each are expressed as follows. The intensities of the incident *s* and *p* waves may be expressed as E_s and E_p, and the intensities of the corresponding reflected waves as R_s and R_p: the absolute phase position δ of each of the incident and reflected *s* and *p* waves may be expressed with its proper subscripts, the ratios are defined as

$$\frac{R_p / R_s}{E_p / E_s} \equiv \tan \psi \text{ and } [\,(\delta p)_r - (\delta s)_r\,] - [\,(\delta p)_j - (\delta s)_j\,] \equiv \Delta \qquad (6.3)$$

then

$$\rho = \tan \psi e^{(j\Delta)} \qquad (6.4)$$

The last equation is the fundamental equation of ellipsometry.

Figure 6.16 describes the geometric manner of conversion of one form of polarized light (linear) to another (elliptical); in general, the incident beam is also elliptically polarized. Actually, linearly polarized light is simply a special case of elliptically polarized light, as is circularly polarized light. Ellipsometry is the technique of measuring changes in the state of polarized light and using these data to determine the complex index of refraction of specimen surfaces. The complex index is composed of two components: real, *n*, and the imaginary, κ. The latter, κ, is related to the absorption coefficient.

Ellipsometers can be used to measure either the complex index of refraction or the thickness of thin films on substrates. In the latter case, if the film is thin enough for light of significant intensity to reach the substrate, the film will alter the apparent complex index of the system from that of the substrate, in proportion to the thickness of the film. A measurement taken with light of one wavelength with a single angle of incidence requires knowledge of the complex index of refraction of both the film material and the substrate material. If the index of refraction of either the film material or the substrate is unknown, measurements must be made with either two different colors of light or at two different angles of incidence. If the index of refraction of neither the film material nor the substrate are known, then measurements must be made with three colors of light or at three angles of incidence, or some combination. Further, if the film consists of two layers, then even more colors of light or angles of incidence must be used. The colors of light and the angles of incidence must be selected with great care for adequate resolving ability of ellipsometry.

7

Magnetic Recording Surfaces

BHARAT BHUSHAN

Contents

7.1 Introduction

Magnetic recording is accomplished by relative motion between a magnetic medium against a stationary (audio and data processing) or rotating (audio, video, and data processing) read/write magnetic head. Under steady operating conditions, a load-carrying air film is formed. There is physical contact between the medium and the head during starting and stopping. The need for increasingly higher recording densities requires that surfaces should be as smooth as possible and that the flying height (physical separation between a head and a medium) be as low as possible. The ultimate objective is to run two surfaces in contact (with practically zero physical separation) if the tribological issues can be resolved. Smooth surfaces lead to an increase in adhesion, friction, and interface temperatures, and closer flying heights lead to occasional rubbing of high asperities and increased wear. Friction and wear issues are resolved by appropriate selection of interface materials and lubricants, by controlling the dynamics of the head and medium, and the environment.

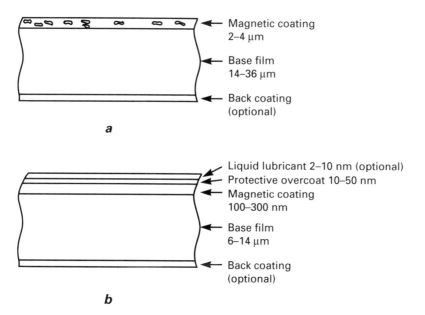

Figure 7.1 Sectional views of (a) particulate and (b) thin-film magnetic tape

7.2 Magnetic Storage Systems

Magnetic storage devices used for information (audio, video, and data processing) storage and retrieval are: tapes, flexible disks, and rigid disk drives.[1] Magnetic media fall into two categories: particulate media, where magnetic particles are dispersed in a polymeric matrix and coated onto a polymer substrate for flexible media (tapes and flexible disks) or onto a rigid substrate (typically aluminum, and more recently glass) for rigid disks; and thin-film media, where continuous films of magnetic materials are deposited onto the substrate by vacuum techniques. For higher recording densities, the trend has been to use thin-film media. Sectional views of tapes and disks are shown in Figures 7.1 and 7.2. Flexible disks are similar in construction to tapes except they have a magnetic coating on both sides and the substrate is generally about 76.2 μm thick.

Particulate magnetic coating is composed of magnetic particles (such as γ-Fe_2O_3, Co-modified γ-Fe_2O_3, CrO_2, metal particles, or barium ferrite), polymeric binders (such as polyester polyurethane or epoxy), and lubricants (mostly fatty acid esters as bulk lubrication for flexible media and perfluoropolyethers as topical lubrication for rigid media). Magnetic thin films used are Co-based alloys, generally deposited by vacuum techniques. A solid overcoat generally made of diamondlike carbon (DLC) or SiO_2 is used for corrosion and wear protection. On top of the solid overcoat, a thin film of perfluoropolyether lubricant is used for low wear.

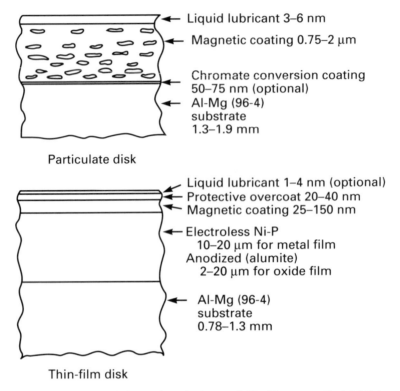

Liquid lubricant 3–6 nm

Magnetic coating 0.75–2 µm

Chromate conversion coating
50–75 nm (optional)

Al-Mg (96-4)
substrate
1.3–1.9 mm

Particulate disk

Liquid lubricant 1–4 nm (optional)
Protective overcoat 20–40 nm
Magnetic coating 25–150 nm

Electroless Ni-P
    10–20 µm for metal film
Anodized (alumite)
    2–20 µm for oxide film

Al-Mg (96-4)
substrate
0.78–1.3 mm

Thin-film disk

Figure 7.2   Sectional views of particulate and thin-film magnetic rigid disk

Magnetic heads used to date are either conventional inductive or thin-film inductive and magnetoresistive (MR) heads.[2] Trends in film-head design have been driven by the desire to capitalize on semiconductor-like technology to reduce fabrication costs. In addition, thin-film technology allows the production of high track density heads with accurate positioning control of the tracks and high reading sensitivity. Conventional heads are a combination of a body forming the air-bearing surface and a magnetic ring core carrying the wound coil with a read-write gap. In the film heads, the core and coils or MR stripes are deposited by thin-film technology.

The air-bearing surfaces of tape heads are cylindrical in shape. The tape is slightly underwrapped over the head surface to generate hydrodynamic lift during read/write operations. The flexible disk heads are either spherically contoured or are flat in shape. The head slider used in rigid disk drives is a two-rail taper-flat design supported by a leaf spring (flexure) suspension made of nonmagnetic steel to allow motion along the vertical, pitch, and roll axes (Figure 7.3). The stiffness of the suspension (about 25 mN mm^{-1}) is several orders of magnitude lower than that of the air bearing (about 0.5 kN mm^{-1}), so that most dynamic variations are taken up by the suspension without degrading the air bearing.

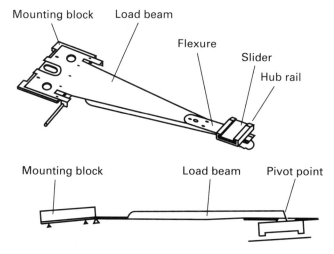

Figure 7.3   IBM 3370/3380/3390-type suspension slider assembly

The air-bearing surfaces of magnetic heads are generally made of hard ceramic materials—Ni-Zn ferrite, Mn-Zn ferrite, calcium titanate (for composite sliders), Al_2O_3-TiC (for thin-film heads), or ZrO_2-Y_2O_3/Al_2O_3-TiC (for thin-film heads). The air-bearing surfaces of some inductive-type heads consist of plasma-sprayed coatings of hard materials such as Al_2O_3-TiO_2 and Y_2O_3-ZrO_2.[1]

7.3 Wear Mechanisms

Head–(Particulate) Tape Interface

The wear of oxide magnetic particles and ceramic head body materials is different from metallic wear because of the inherent brittleness and the relatively low surface energy of ceramics. The first sign of ferrite head wear with a magnetic tape is the appearance of very small scratches on the head surface (Figure 7.4a). The physical scale of scratches is usually very fine, and scratches as small as 25 nm have been reported.[1] Ferrite surface is microscopically removed in a brittle manner as stripes or islands, depending on the smoothness of the tape surface. Wear generally occurs by microfragmentation of the oxide crystals in the ceramic surface. Fragmentation is the result of cleavage and transgranular fractures, dominated by intergranular fracture. We note that the worn head surface is work hardened, which reflects a shift from a mechanism dominated by transgranular fracture rather than dominated by intergranular fracture. Figure 7.4b shows a region where fracture and rupture have occurred. Such regions are commonly called *pullouts*. Debris originating from these regions causes additional small-scale plastic deformation and grooves

a   b

Figure 7.4   **SEM micrographs of worn Ni-Zn ferrite head with a CrO$_2$ tape A: a) abrasion marks (direction of tape motion-left to right) and b) surface pullout**

(three-body abrasion). We also note that the ferrite head surface is work hardened with a large compressive stress field after wear, which is detrimental to magnetic signal amplitude.

Head wear depends on the physical properties of head and medium materials, drive operating parameters, and environmental conditions. A linear relationship between wear rate and material hardness has been reported.[1] Head wear also depends on the grain size of the head material, magnetic particles, tape-surface roughness, isolated asperities on the tape surface, tape tension, and tape sliding speed.[1] An increase in head wear with an increase in the surface roughness or number of isolated asperities of the tape surface has been reported.[1] Wear rate also increases with humidity, above about 40–60% relative humidity.[1] An increase in abrasive wear at high humidities is believed to be due to moisture-assisted fracture (or static fatigue) of the grains to yield finer particles.

After usage, head sliders sometimes become coated with thin layers of a new organic material of high molecular weight called a *friction polymer* or *tribopolymer*.[1] Friction is essential for the formation of these materials. Another requirement for the formation of friction polymers in a rubbing contact is that one of the surfaces, lubricant, or even a material nearby should be organic. It seems clear that all friction polymers are products of chemical reaction, whether they derive initially from solid polymers, or from organic liquids or vapors. Friction polymers are found on head surfaces. These result in discoloration of the head surface, giving an appearance of brown or blue color, and therefore are sometimes called brown or blue stains, respectively.

During contact of particulate tape with the head in start and stops, or during partial contact in streaming, binder and magnetic particle debris is generated, primarily by adhesive wear mode. The debris can be either loose or adherent. Tape debris, loose magnetic particles, worn head material, or foreign contaminants are introduced between the sliding surfaces and abrade material off each. The debris that adheres to drive components leads to polymer–polymer contact, whose friction is

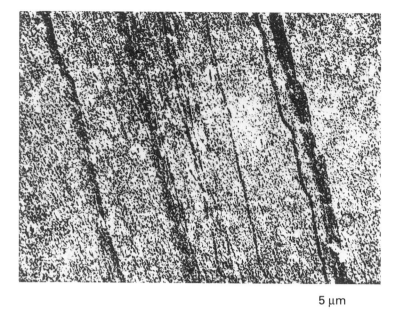

5 μm

Figure 7.5 SEM micrograph of a worn particulate disk surface after extended use in contact-start-stop against Al_2O_3–TiC slider

higher than that of rigid material–polymer contact and can lead to magnetic errors and sometimes to catastrophic failures.[1]

Head–(Particulate) Rigid Disk Interface

During asperity contacts, disk debris can be generated by adhesive, abrasive, and impact wear. The wear debris, generated during the manufacturing process (burnishing or buffing) of the disk, and foreign contaminants present can be trapped at the interface, which results in three-body abrasive wear, or can be transferred to the slider, resulting in performance degradation of the air bearing. The flash temperatures generated at asperity contacts can render any boundary lubricants ineffective and can degrade the mechanical properties of the disk binder, increasing the real area of contact; this results in high friction and high disk wear. Any of these mechanisms can lead to head crash. The environment (humidity and temperature) has a significant effect on the head–disk friction and wear or debris generation.

Microscopic observations of disk and head surfaces after head crash in a contact-start-stop (CSS) test show circumferential wear grooves, and support either of the two-body or three-body abrasive wear modes (Figure 7.5). Novotny et al.[3, 4] reported the complete removal of lubricant from the start–stop track at least in one region, and degradation of lubricant preceded the final stage of disk failure.

Lubrication of the disks increased the number of sliding cycles until the coating began to wear through by up to 1000 times. The number of sliding cycles until frictional failure occurred was proportional to the areal density of lubricant.

Head–(Thin-Film) Rigid Disk Interface

Magnetic films used in the construction of thin-film disks are soft and have poorer wear resistance than the particulate disks which are loaded with hard magnetic particles and load-bearing alumina particles. Thin-film disks have smoother surfaces and are designed to fly at a lower height than particulate rigid disks, which results in higher friction and an increased potential of head–disk interactions. During normal drive operation, the isolated asperity contacts of head and disk surfaces result in wearing out the disk surfaces by adhesive and impact wear modes and generating debris. In addition, any asperity contacts would result in the maximum shear stress at the disk subsurface, which may initiate a crack. Repeated contacts would result in crack propagation (subsurface fatigue), leading to delamination of the overcoat and the magnetic layer. Isolated contacts in a clean environment generate very fine wear debris (primarily made of magnetic film and overcoat) which subsequently results in rather *uniform disk wear* from light burnishing (three-body abrasion). The disk wear causes high friction and wear in subsequent contacts, resulting in head crash.[1]

External contamination abrades the overcoat readily and results in localized damage of the disk surface by three-body abrasion. Wear debris generated at the interface invades the spacing between the head slider and the disk or transfers to the head slider, making the head slider unstable, which leads to additional debris and results in head crash in both the start–stop and flying modes.

Engel and Bhushan[5] have developed a head–disk interface failure model for thin-film disks. The principal physical variables include the sliding speed, surface topography, mechanical properties, coefficient of friction, and wear rate. Surface protrusions, such as asperities and debris particles, induce impact and sliding encounters, which represent a damage rate. Failure occurs when a specific damage rate, which is characteristic for the system, is reached. Modeling uses a set of topographic parameters describing the changing, wearing surface.

Marchon et al.[6] and Strom et al.[7] studied the wear behavior of unlubricated thin-film disk with a carbon overcoat sliding against ceramic sliders in various environments. Strom et al.[7] tested unlubricated carbon-coated disks (disks B_1, with no lubricant) and ZrO_2-Y_2O_3-coated disks (disks B_2 with no lubricant) in a sliding test against commercial Al_2O_3-TiC slider (Figures 7.6). The average friction of the carbon-coated disks increased smoothly only in the oxygen environments which indicates poor durability. At sustained high friction, debris was generated which reduced the real area of contact, and the friction dropped. In the case of zirconia-coated disk, no consistent difference in the tribological behavior between various gas environments was observed.

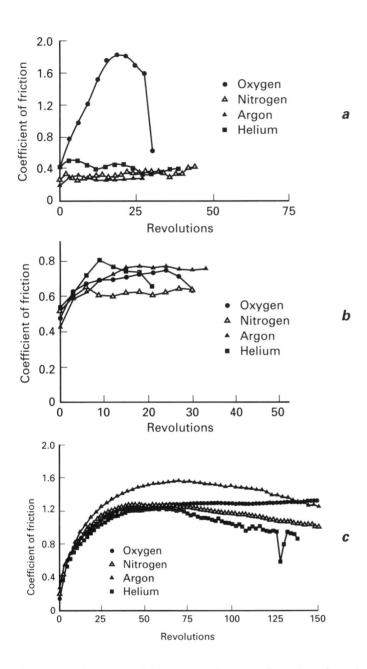

Figure 7.6   Coefficient of friction as a function of number of revolutions during sliding of Al_2O_3-TiC slider against an unlubricated thin-film disk at a normal load of 150 mN and a sliding speed of 60 mm/s: (a) disk with carbon overcoat (B_1, with no lubricant) in dry gases, (b) disk with zirconia overcoat (B_2, with no lubricant) in dry gases, and (c) disk with carbon overcoat (B_1, with no lubricant) in various gases, all at 4% RH [7]

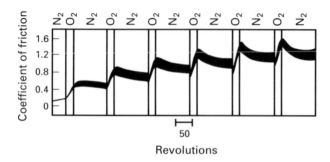

Figure 7.7   Coefficient of friction as a function of number of revolutions during sliding of Mn-Zn ferrite slider against an unlubricated thin-film disk at a normal load of 100 mN and a sliding speed of 60 mm/s in various dry gases[6]

The sliding test on carbon-coated disks was also conducted in the presence of humid gases. In this case, little difference in the friction behavior for different gases was observed (Figure 7.6c). In all gases, the coefficient of friction increased smoothly to about 1.4 during the course of about 50 revolutions. The increase in friction in the oxygen or humid environment for the carbon-coated disk can be explained by the oxidation of the carbon film under rubbing, a tribochemical reaction. In the case of sliding on the zirconia-coated disk, no material was removed through oxidation. Wear occurs through mechanical means only, regardless of the concentration of oxygen in environment. Lubricated carbon disks were less sensitive to the concentration of oxygen in the environment.

Marchon et al.[6] have also reported carbon oxidation as a key process that contributes significantly to wear and friction increase with test cycles. In sliding experiments with Mn–Zn ferrite or calcium titanate sliders and unlubricated thin-film disks with carbon overcoats, Marchon et al. reported that there is a gradual increase in the coefficient of friction with repeated sliding contacts performed in air; however, in pure nitrogen, no friction increase is observed, the coefficient of friction remaining constant at 0.2. The alternate introduction of oxygen and nitrogen elegantly showed the role of these gases (Figure 7.7). Contact start–stop tests exhibited same the effect. Marchon et al. suggested that the wear process in the oxygen environment involved oxygen chemisorption on the carbon surface and a gradual loss of carbon through the formation of CO/CO_2 due to the action of the slider.

7.4 Lubrication Mechanisms

Mechanical interactions between the head and the medium are minimized by the lubrication of the magnetic medium. The primary function of the lubricant is to reduce the wear of the magnetic medium and to ensure that friction remains low throughout the operation of the drive. The main challenge, though, in selecting the best candidate for a specific surface is to find a material that provides an acceptable wear protection for the entire life of the product, which can be several years in duration. There are many requirements that a lubricant must satisfy in order to guarantee an acceptable life performance. An optimum lubricant thickness is one of these requirements. If the lubricant film is too thick, excessive stiction and mechanical failure of the head–disk is observed. On the other hand, if the film is too thin, protection of the interface is compromised, and high friction and excessive wear will result in catastrophic failure. An acceptable lubricant must exhibit properties such as chemical inertness, low volatility, high thermal, oxidative and hydrolytic stability, shear stability, and good affinity to the magnetic medium surface.

Fatty acid esters are excellent boundary lubricants, and esters such as tridecyl stearate, butyl stearate, butyl palmitate, butyl myristate, stearic acid, and myrstic acid are commonly used as internal lubricants, roughly 1–3% by weight of the magnetic coating, in tapes and flexible disks. The fatty acids involved include those with acid groups with an even number of carbon atoms between C_{12} and C_{22}, with alcohols ranging from C_3 to C_{13}. These acids are all solids with melting points above the normal surface operating temperature of the magnetic media. This suggests that the decomposition products of the ester via lubrication chemistry during a head–flexible medium contact may be the key to lubrication.

Topical lubrication is used to reduce the wear of rigid disks. Perfluoropolyethers (PFPEs) are chemically the most stable lubricants with some boundary lubrication capability, and are most commonly used for topical lubrication of rigid disks.[1] PFPEs commonly used include Fomblin Z and Fomblin Y lubricants, made by Montiedison, Italy; Krytox 143 AD, made by Dupont, USA; and Demnum, made by Diakin, Japan; and their difunctional derivatives containing various reactive end groups, e.g., hydroxyl (Fomblin Z-DOL), piperonyl (Fomblin AM 2001), and iso-cyanate (Fomblin Z-DISOC), all manufactured by Montiedison. The difunctional derivatives are referred to as reactive (polar) fluoroether lubricants. The chemical structures, molecular weights, and viscosities of various types of PFPE lubricants are given in Table 7.1. We note that rheological properties of thin-films of lubricants are expected to be different from their bulk properties.[8] Fomblin Z is a linear PFPE, and Fomblin Y and Krytox 143 AD are branched PFPE, where the regularity of the chain is perturbed by $-CF_3$ side groups. The bulk viscosity of Fomblin Y and Krytox 143 AD is almost an order of magnitude higher than the Z type. The molecular coil thickness is about 0.8 nm for these lubricant molecules. The monolayer thickness of these molecules depends on the molecular conformations of the polymer chain on the surface.

Fomblin Y and Z are most commonly used for particulate and thin-film rigid disks. Usually, lubricants with lower viscosity (such as Fomblin Z types) are used in thin-film disks in order to minimize stiction.

Lubricant	Formula	Molecular weight, Daltons	Kinematic viscosity, c St(mm^2/s)
Fomblin Z-25	$CF_3-O-(CF_2-CF_2-O)_m-(CF_2-O)_n-CF_3$	12800	250
Fomblin Z-15	$CF_3-O-(CF_2-CF_2-O)_m-(CF_2-O)_n-CF_3$ (m/n ~ 2/3)	9100	150
Fomblin Z-03	$CF_3-O-(CF_2-CF_2-O)_m-(CF_2-O)_n-CF_3$	3600	30
Fomblin Z-DOL	$HO-CH_2-CF_2-O-(CF_2-CF_2-O)_m-(CF_2-O)_n-CF_2-CH_2-OH$	2000	80
Fomblin AM2001	Piperonyl-$O-CH_2-CF_2-O-(CF_2-CF_2-O)_m-(CF_2-O)_n-CF_2-CH_2-O$-piperonyl*	2300	80
Fomblin Z-DISOC	$O-CN-C_6H_3-(CH_3)-NH-CO-CF_2-O-(CF_2-CF_2-O)_n-(CF_2-O)_m-CF_2-CO-NH-C_6H_3-(CH_3)-N-CO$	1500	160
Fomblin YR	$CF_3-O-(\underset{F}{\overset{CF_3}{C}}-CF_2-O)_m-(CF_2-O)_n-CF_3$ (m/n ~ 40/1)	6800	1600
Demnum S-100	$CF_3-CF_2-CF_2-O-(CF_2-CF_2-CF_2-O)_m-CF_2-CF_3$	5600	250
Krytox 143AD	$CF_3-CF_2-CF_2-O-(\underset{F}{\overset{CF_3}{C}}-CF_2-O)_m-CF_2-CF_3$	2600	—

Table 7.1   Chemical structure, molecular weight, and viscosity of perfluoropolyether lubricants

*3,4 – methylenedioxybenzyl

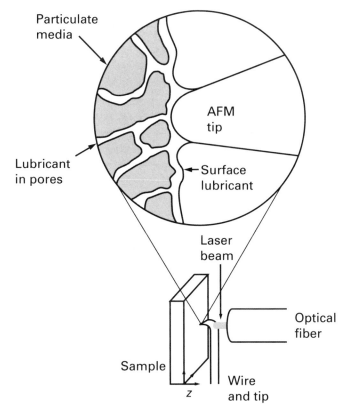

Particulate
media

Lubricant
in pores

AFM
tip

Surface
lubricant

Laser
beam

Optical
fiber

Sample

Wire
and tip

z

Figure 7.8   **AFM (bottom) and the enlarged region and cross sectional**
view (top) of the particulate disk surface and the AFM tip
during lubricant film thickness measurement[8]

Measurement of Localized Lubricant Film Thickness

The local lubricant thickness is measured by Fourier transform infrared spectroscopy (FTIR), ellipsometry, angle-resolved X-ray photon spectroscopy (XPS), scanning tunneling microscopy (STM), and atomic force microscopy (AFM).[9–11]
Ellipsometry and angle-resolved XPS have excellent vertical resolution, on the order
of 0.1 nm, but their lateral resolutions are on the order of 1 μm and 0.2 mm, respectively. STM and AFM can measure the thickness of the liquid film with a lateral resolution on the order of the tip radius, about 100 nm, which it is not possible to
achieve by other techniques.

The schematic of AFM used for measurement for localized lubricant-film thickness by Mate et al.[10] is shown in Figure 7.8. The lubricant thickness is obtained by
measuring the forces on the tip as it approaches, contacts, and pushes through the
liquid film. In the top part of Figure 7.8 is a diagram of an AFM tip interacting
with a lubricant-covered particulate disk. A typical force-versus-distance curve for a

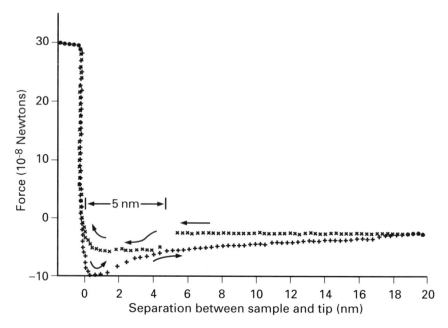

Figure 7.9 The force perpendicular to the surface acting on the tip as a function of particulate disk A sample position, as the sample is first brought into contact with the tip (x) and then pulled away (+). The sample is moved with a velocity of 100 nm/s, and the zero sample position is defined to be the position where the force on the tip is zero when in contact with the sample. A negative force indicates an attractive force

tungsten tip of radius ~100 nm dipped into a disk surface coated with a perfluorinated polyether lubricant is shown in Figure 7.9. As the surface approaches the tip, the liquid wicks up, causing a sharp onset of attractive force. The so-called meniscus force experienced by the tip is ~$4\pi r\gamma$ where r is the radius of the tip and γ is the surface tension of the liquid. In Figure 7.9, the attractive force measured is about 5×10^{-8} Newtons. While in the liquid film, the forces on the lever remain constant until repulsive contact with the disk surface occurs. The distance between the sharp snap-in at the liquid surface and the *hard-wall* of the substrate is proportional to the lubricant thickness at that point. (The measured thickness is about 2 nm larger than the actual thickness due to a thin layer of lubricant wetting the tip.[10]) When the sample is withdrawn, the forces on the tip slowly decrease to zero as a long meniscus of liquid is drawn out from the surface.

Particulate disks were mapped by Mate et al.[10] and Bhushan and Blackman.[11] The distribution of lubricant across an asperity was mapped by collecting force-versus-distance curves with the AFM in a line across the surface. Disk *A* was coated with nominally 20–30 nm of lubricant, but by AFM the average thickness is 2.6 ±1.2 nm. (The large standard deviation reflects the huge variation of lubricant thickness across the disk.) A large percentage of the lubricant is expected to reside

MAGNETIC RECORDING SURFACES Chapter 7

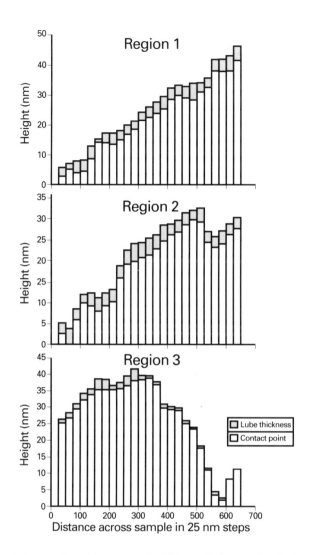

**Figure 7.10   Histograms of lubricant thickness distribution
at three different regions of particulate disk _A_**

below the surface in the pores. In Figure 7.10, we show histograms of lubricant thickness across three regions on disk _A_. The light part of the bar represents the hard wall of the substrate and the dark part on the top is the thickness of the lubricant. Each point on the histogram is from a single force-versus-distance measurement, with points separated by 25-nm steps. The lubricant is not evenly distributed across the surface. In regions 1 and 2 there is more than twice as much lubricant than there is on the asperity (region 3). There are some points on the top and the side of the asperity which have no lubricant coating at all.[11]

Lubricant–Disk Surface Interactions

The adsorption of the lubricant molecules is due to van der Waals forces, which are too weak to offset the spin-off losses, or to arrest displacement of the lubricant by water or other ambient contaminants. Considering that these lubricating films are on the order of a monolayer thick and are required to function satisfactorily for the duration of several years, the task of developing a workable interface is quite formidable.

An approach aiming at alleviating these shortcomings is to enhance the attachment of the molecules to the overcoat, which, for most cases, is sputtered carbon. There are basically two approaches which have been shown to be successful in bonding the monolayer to the carbon. The first relies on exposure of the disk lubricated with neutral PFPE to various forms of radiation, such as low-energy X ray,[12] nitrogen plasma,[13] or far ultraviolet (e.g., 185 nm).[14] Another approach is to use chemically active PFPE molecules, where the various functional (reactive) end groups offer the opportunity of strong attachments to specific interface. These functional groups can react with surfaces and bond the lubricant to the disk surface, which reduces its loss due to spin off and evaporation. Their main advantage, however, is their ability to enhance durability without the problem of stiction usually associated with weakly bonded lubricants.[1] The effect of bonded lubricant was demonstrated elegantly in recent AFM experiments by Blackman et al.[15] They found that when a AFM tip is brought into contact with a molecularly thin film of a nonreactive lubricant, a sudden jump into adhesive contact is observed. The adhesion was initiated by the formation of a lubricant meniscus surrounding the tip pulling the surfaces together by Laplace pressure. However, when the tip was brought into contact with a lubricant film which was firmly bonded to the surface, only a marginal adhesion, mostly due to van der Waals forces, was measured (Figure 7.11).

Lubricant Degradation

Contacts between the slider and the lubricated disk lead to lubricant loss (Figure 7.12).[3, 4, 16] The lubricant polymer chain is scissioned during slider–disk contacts. The transient interface temperatures may be high enough[17] to lead to the direct evaporation or desorption of the original lubricant molecules.

Novotny and Karis[3] studied the difference between mechanisms by comparison of the Fomblin Y and Z lubricants with different chemical structure but approximately the same molecular weight:

1 In sliding, Y is removed from the surface more rapidly than Z.
2 During flying, Z is removed more rapidly from the surface than Y.
3 Z is thermally decomposed more rapidly than Y.[18]
4 Migration rates of Z are higher than those of Y.

Polymer chain scission can be driven by mechanical, triboelectric, or thermal mechanisms.[18, 19] The lubricant thickness is typically 1–5 nm, and at 1–5 m/s

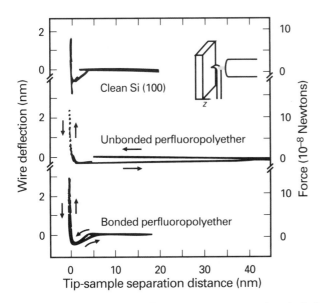

Figure 7.11 Force–versus–distance curves comparing the behavior of clean Si(100) surface to a surface lubricated with free and nonbonded PFPE lubricant, and a surface where the PFPE lubricant film was thermally bonded to the surface[15]

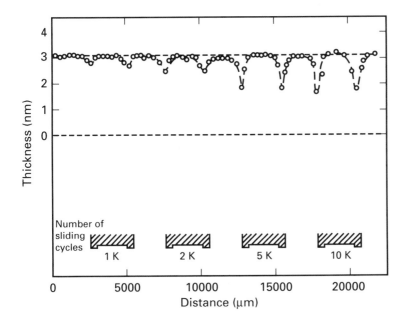

Figure 7.12 XPS cross-sectional profiles of lubricant thickness immediately after sliding of an Al_2O_3-TiC slider on a carbon-coated thin-film disk lubricated with 3 nm of Fomblin Y with a load of 150 mN and a sliding velocity of 1 m/s[3]

sliding velocity, the shear rate is 0.2×10^9/sec to 5×10^9sec. At such high shear rates, there can be much energy imparted to the polymer chain, inducing the chains to slide over one another. If the molecules are strongly interacting with the surface, the sliding may be hindered. Additional hindrance to interchain sliding can be the presence of a bulky side group, such as the $-CF_3$ group on the Y lubricant. The additional intermolecular friction of the chains with the side group can hinder the rapid configurational adjustments required to support such high shear rates. The chains can be broken by tearing bonds apart along the polymer backbone and reduced to volatile products which provide the route for the observed lubricant loss when contacts occur in sliding or flying. It also follows that the loss of Y with the side group should be more rapid than that of the linear Z lubricant.

Moreover, potential differences up to 0.1 V and 0.5 V are measured on thin-film and particulate media surfaces, respectively, between areas on and off the sliding tracks. Corresponding electric fields for slider-disk separations lead to alteration of organic materials when the localized currents pass between the disk and slider asperities.

Thermal decomposition of perfluoropolyethers can proceed by a free radical mechanism which involves initiation, propagation, and termination. The relative rates of initiation and propagation should be different for the two lubricants because of their chemical structure. From thermogravimetric analysis, for an equivalent rate of thermal decomposition on iron oxide, the temperature of Y must be held about 100°C higher than that of Z. However, the loss of lubricant by thermal decomposition may also depend on the relative displacement and migration rates. Thus, the higher loss rate of Y than Z in sliding can also be consistent with the thermal decomposition pathway. Faster degradation rates for Z lubricants than for Y lubricants are believed to be due to the catalytic effect of high concentration of acetal units (CF_2-O) in Z lubricants which results in chain scission.[18,19]

In flying, there is a displacement of lubricant to the outside of the track and a decrease of lubricant in the track. The displacement can be attributed to the repetitive application of pressure in the slider air bearing and intermittent contacts between the slider rails and the disk. Taking into account both the displacement and decrease in the lubricant level during flying, there is a net loss of lubricant from the disk surface. One possible mechanism for the loss is by aerosol droplet formation. This mechanism is reasonable given the tremendous negative pressure gradient at the trailing edge of the slider. The typical pressure increase under the air-bearing rail is about 10^5 Pa, and this pressure drop occurs over about 10 μm, yielding a trailing edge pressure gradient of 10^{10} Pa/m. The rate of aerosol generation can depend on the lubricant surface tension (which is about the same for Z and Y), and the lubricant molecular configuration (effectively a flow property). The $-CF_3$ side group can act to hinder chain slippage (flow) required for efficient aerosol generation, lowering the loss rate of Y below that of Z in flying.

References

1 B. Bhushan, *Tribology and Mechanics of Magnetic Storage Devices*, Springer Verlag, New York, 1990.

2 C.D. Mee and E.D. Daniel, Eds., *Magnetic Recording Handbook*, McGraw-Hill, New York, 1990.

3 V.J. Novotny and T.E. Karis, *Adv. Info. Storage Syst.*, **2**, 137–152, 1991.

4 V.J. Novotny, T.E. Karis, and N.W. Johnson, *J. Trib.*, Trans. ASME, **114**, 61–67, 1992.

5 P.A. Engel and B. Bhushan, *J. Trib., Trans. ASME*, **112**, 299–303, 1990.

6 B. Marchon, N. Heiman, and M.R. Khan, *IEEE Trans. Magn.*, **26**, 168–170, 1990.

7 B.D. Strom, D.B. Bogy, C.S. Bhatia, and B. Bhushan, *J. Trib., Trans. ASME*, **113**, 689–693, 1991.

8 J.N. Israelachvili, P.M. McGuiggan, and A.M. Homola, *Science*, **240**, 189–191, 1988.

9 Y. Kimachi, F. Yoshimura, M. Hoshino, and A. Terada, *IEEE Trans. Magn.*, **Mag-23**, 2392–2394, 1987.

10 C.M. Mate, M.R. Lorenz, and V.J. Novotny, *IEEE Trans. Magn.*, **Mag-26**, 1225–1228, 1990.

11 B. Bhushan and G.S. Blackman, *J. Trib., Trans. ASME*, **113**, 452–457, 1991.

12 R. Heidemann and M. Wirth, *IBM Tech. Disclosure Bull.*, **27**, 3199–3205, 1984.

13 A.M. Homola, L.J. Lin, and D.D. Saperstein, US Patent 4,960,609, Oct. 2, 1990.

14 D.D. Saperstein and L.J. Lin, *Langmuir*, **6**, 1522–1524, 1990.

15 G.S. Blackman, C.M. Mate, and M.R. Philpott, *Phys. Rev. Lett.*, **65**, 2270–2273, 1990.

16 Y. Hu and F.E. Talke, *Tribology and Mechanics of Magnetic Storage Systems*, Vol. 5 (B. Bhushan and N.S. Eiss, Eds.), **SP-25**, 43–48, STLE, Park Ridge, IL, 1988.

17 B. Bhushan, *J. Trib., Trans. ASME*, **114**, 420–430, 1992.

18 P.H. Kasai, W.T. Tang, and P. Wheeler, *Appl. Surf. Sci.*, **51**, 201, 1991.

19 P.H. Kasai, *Adv. Info. Storage Syst.*, **4**, 1992.

8

Surface Analysis of Bearings

MICHAEL R. HILTON and STEPHEN V. DIDZIULIS

Contents

8.1 Introduction

Precision ball bearings used in gyroscopes and in spacecraft pointing mechanisms critically depend upon superior lubricant performance in order to function properly. The lubricants are solids, liquids, or greases, with the latter two usually containing boundary or extreme pressure additives. These additives modify the bearing surface composition, which reduces wear, i.e., provides lubrication during low-speed operation. These mechanisms generally contain the optimum amount of lubricant needed to minimize torque—either insufficient or excessive quantities of lubricant lead to an increase in torque. Thus, bearing failure for these mechanisms generally occurs because the lubricant becomes inadequate, through chemical degradation, evaporative loss, or migration of the lubricant away from the contact zone. In contrast, bearing failure by classical fatigue occurs in only 10% of steel aerospace bearings and is virtually nonexistent in spacecraft bearings.[1, 2] Therefore, the failure analyses of these precision bearings focus on the determination of lubricant failure and the measurement of bearing wear. This chapter reviews the techniques used to conduct a failure analysis of precision bearings. The review is arranged in a manner consistent with the sequence of bearing disassembly. The information to be gained by and the limitations of each technique will be identified, and appropriate examples will be presented. The reader should note that a multitechnique approach is critical to determining the cause of bearing failure. Hopple has written an excellent guide to the problem with regard to lubricant analysis and microexamination.[3] The present text reviews those techniques, but also emphasizes their integration with surface analysis. Further details of the surface analysis techniques can be found in other volumes of this series or in the literature.[4]

8.2 Disassembly

Examination, Optical Microscopy, and Photography

Physical examination of the bearing provides the initial indication of the failure mode, and in some cases can suggest a particular emphasis for the application of subsequent techniques. The following should be noted by low-power (10×–100×) optical microscopy of the bearing: condition of lubricant (discolored, dry, presence or absence, etc.), presence and nature of debris, retainer condition (fractured, discolored, etc.), and any other features of interest. Photography provides a visual record of the bearing condition. Photographs of the disassembled bearing can also act as a guide to direct the location of further analyses. Most bearings can be disassembled using conventional methods. For less accessible components that are cylindrical in shape, disassembly using a jeweler's lathe is preferred. Lathe turnings, unlike grinding or sawing debris, are easily distinguishable from most forms of bearing wear or contamination particles.[3]

Gas Analysis by Mass Spectrometry

Some mechanisms, particularly gyroscopes, have bearings enclosed in hermetically sealed atmospheres to prevent contamination. The sealed atmosphere can be sampled and analyzed by mass spectrometry (residual gas analyzer) to verify the integrity of the seal and to detect impurities. For space applications, many bearing qualification tests are conducted in vacuum to simulate the operational environment. A mass spectrometer can be attached to the vacuum test chamber to ensure that a clean vacuum (e.g., no leaks from atmosphere, no cleaning solvents present) is maintained during the test.

Lubricant Analysis and Removal

When bearings are lubricated by solids, such as MoS_2 or polymers, these lubricant films can be analyzed along with the bearing surfaces as discussed in Sections 8.3 and 8.4. However, liquid lubricants and the liquid portions of greases are not amenable to those techniques. Special techniques must be employed to analyze these lubricants prior to microexamination and surface analysis of the bearings. Indeed, the liquids must be removed prior to bearing analysis, and the removal needs to be executed in a manner that minimizes disturbance to the bearing surface. Greases, which consist of liquid lubricant and thickener (often soap compound) components, require special procedures for separation and independent analyses of these components. Detailed procedures for lubricant analyses have been reported by Carré[5] and Hopple,[3] but aspects will be briefly reviewed here. Fluid lubricants are analyzed for the following reasons: to verify that the appropriate (specified) lubricant was used; to assess chemical degradation (polymerization, fragmentation, oxidation, etc.) of the lubricant; to determine if evaporative loss of the lighter molecular weight fractions has occurred; to detect fluid contamination; to select an appropriate solvent for bearing cleaning prior to surface analysis; and to determine the extent of grease separation (oil loss).

Two common techniques for lubricant analysis are infrared spectroscopy (IR) and gas chromatography (GC). IR measures the vibrational modes of organic molecules, which can be used to assess functional groups (e.g., alcohols and ketones) of the lubricant. GC separates the lubricant molecules by boiling point, providing an assessment of the molecular weight distribution, which in part determines the viscosity of the liquid lubricant. In some GC systems, special columns can be used to separate functional groups. If the GC is combined with a mass spectrometer (GC–MS), the molecular structure of the separated molecules can be further assessed. In all cases, careful comparison of data from bearing lubricant samples, and from lubricant and other standards is required to meet the objectives mentioned in the previous paragraph.

Solid debris particles may be present in the liquid and can be collected by filtration and investigated with the techniques of Sections 8.3 and 8.4. The thickener component of greases can be preserved by freeze-drying, and its structure examined by microscopy. For example, overheating of a grease can lead to melting of the thickener, which causes the porous structure to collapse, rendering it unable to retain sufficient oil.[3]

Prior to microscopy and surface analysis in vacuum, liquid lubricants have to be removed. Heptane, which can be obtained readily in high purity because it is used as a standard for GCs, is often used to remove hydrocarbon liquids, while Freon is utilized to remove fluorocarbon liquids. Although ultrasonic agitation is generally used, we try to limit the duration to the minimum time sufficient to remove the liquid without further mechanically disturbing remnant additive-derived surface films. For this reason, bearing balls are often separated to prevent impingement by rolling during cleaning.

8.3 Microexamination

Scanning Electron Microscopy

Scanning electron microscopy (SEM) is one of the more powerful tools for bearing failure analysis. SEM provides direct assessment of the bearing surface morphology so that wear can be identified. Excellent texts of SEM studies of wear and fatigue exist and should be consulted.[6, 7] In the present context, SEM also is very useful in studying remnant lubricant films, remnant additive films, debris particles, or the freeze-dried thickener components of greases.[8] SEM provides structural information on the morphology of these surface regions. In combination with X-ray spectroscopy, either using energy-dispersive (EDS) or wavelength-dispersive (WDS) detectors, SEM can provide the spatially resolved composition distribution of elements present near the surface (0.5–1 μm depth).

A key advantage in using SEM to complement the surface analysis techniques in Section 8.4 is that SEM, because of high signal collection rates and less-stringent vacuum requirements, enables more rapid analysis of bearing surfaces compared to the ultrahigh vacuum (UHV) techniques. SEM is used to survey the various types of surface regions present on bearing surfaces. Representative regions are then selected for further, more time-consuming surface analysis. SEM also has better spatial resolution than the surface spectroscopies. SEMs that can produce electron beam diameters (spot sizes) of 4–6 nm are now available.

In examining bearing surfaces, multiple-mode images (i.e., secondary electron (SE) versus backscattered electron (BSE), and high-energy versus low-energy imaging modes) of surface regions are particularly informative. The contrast in SE images is strongly dependent upon surface topography relative to the detector, but is relatively insensitive to the composition (atomic number Z) of the irradiated region. BSE images are strongly dependent upon atomic number, as well as topography, with higher Z elements generating more backscattered electrons. Lower primary-beam energy in the BSE mode enhances the surface sensitivity of the image, emphasizing elements at the surface, although spatial resolution degrades because the electron spot size increases as the beam energy decreases.

For example, bearing raceways, operated under low-speed oscillatory conditions and lubricated with a synthetic hydrocarbon oil, a multiply-alkylated cyclopentane (MAC), and the extreme pressure additive lead naphthenate (Pbnp) were studied by SEM. The BSE mode at 5 keV revealed the presence of a bright overlayer that was less evident at 20 keV and not evident in the SE mode, as shown in Figure 8.1. Surface analysis (Section 8.4) revealed that the bright regions contained Pb and C

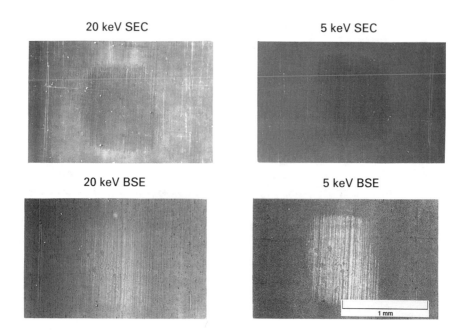

| 20 keV SEC | 5 keV SEC |
| 20 keV BSE | 5 keV BSE |

Figure 8.1   SEM photographs showing a bearing raceway lubricated with MAC oil and lead naphthenate (Pbnp). A Pb-containing overlayer film in the contact region is very evident (i.e., has a very bright appearance) using BSE mode, which is more surface sensitive (particularly at 5 keV)

decomposed from Pbnp. The high atomic number of Pb (82) caused the bright contrast in BSE mode, relative to Fe (26) or C (6). These Pb-containing films provided protection against microdeformation and wear that occurred in tests of the oil without Pbnp additive, as shown in Figure 8.2.

The BSE mode of SEM was also useful in studying the wear and deformation of sputter-deposited MoS_2 solid-lubricant films on thrust bearings operated unidirectionally. Figure 8.3 shows a bearing after failure (defined as unacceptable motor torque). A 1-μm film was initially present on the bearing. The micrographs show the pathway of ball contact where most of the film has been removed, as confirmed by EDS and surface analysis (Section 8.4).

Although the BSE mode can be useful in spatially resolving different compositions, there are limitations. Topography changes also affect contrast in BSE. Correlation of BSE gray levels with different chemical phases has to be continually verified by EDS/WDS, or by surface analysis, during investigation. New samples require special scrutiny to confirm previous trends. Contrast in the BSE mode is very system dependent. Although Pb yields strong contrast in BSE, P from the additive tricresyl phosphate (TCP) does not yield a strong contrast. BSE imaging provides little benefit in that case.

With Pbnp   Without Pbnp

Figure 8.2   SEM micrographs comparing two bearing raceways lubricated with a MAC oil. One raceway had Pbnp present, as in Figure 8.1, while the other raceway was lubricated with only the oil. Deformation, evident with the pure oil, is avoided when the Pbnp additive is present. The SEM micrographs provide information on the bearing and lubricant overlayer surfaces

Profilometry

Profilometry by stylus methods is used to measure bearing surface roughness and curvature. Use of this technique can determine if wear or additive film build-up has occurred in contact regions. The wear or redistribution of solid lubricants in bearings can be studied. (Initial solid-lubricant film thickness prior to testing can also be measured if sharp steps are present between coated and uncoated regions. This is usually done on witness plates, as opposed to bearings.) The stylus tip may deform the soft solid lubricant or additive film and yield an erroneously low thickness. If the stylus deformation width on the film is examined and measured by SEM, the depression and correct thickness can be calculated because the tip geometry is known.[9]

8.4 Surface Analysis

Auger Electron Spectroscopy

Auger electron spectroscopy (AES), often executed in a scanning Auger microprobe (SAM), is used in the analysis of bearings to determine the elemental composition of

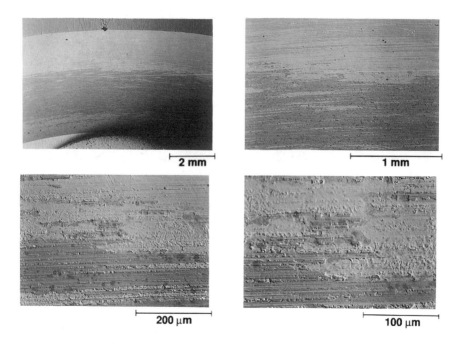

Figure 8.3   SEM micrographs of a thrust-bearing raceway lubricated with sputter-deposited MoS$_2$ (1-μm thick) after testing and failure, defined as a significant torque rise. The BSE mode reveals that the film has been significantly removed from the contact region, appearing as dark-gray areas in these BSE micrographs. Note the presence of carbides in the region where film has been removed

the near-surface region (< 10 nm), including overlayers, contaminants, and exposed bearing material. If argon ion sputtering is used to remove material, compositional depth profiles can be obtained, which can help to determine the thickness of overlayers and contaminants. Because the electron beam used to generate Auger electrons can be focused and scanned, in principle, down to a beam size of 30 nm, spatially resolved maps of surface composition can be generated. In practice, SAM is used after SEM investigation has identified representative regions for analysis. (The electron beam of the SEM might, in some cases, damage a carbonaceous surface layer or induce contamination upon the surface. The investigator might elect to examine some samples directly with AES (or XPS) after optical examination.) When detecting secondary electrons the SAM, can be operated as a SEM, although only the most recent instruments have beam sizes, and therefore spatial resolution, comparable to standard SEMs. The minimum detection limit for most elements by AES is 0.1–1 %at. Older AES systems, many of which are still in use, display peak derivatives because lock-in amplifiers are used in the detection scheme. The derivative peak height is normally related to the quantity of element present in the detected volume, after normalization and cross section sensitivities are taken into account.

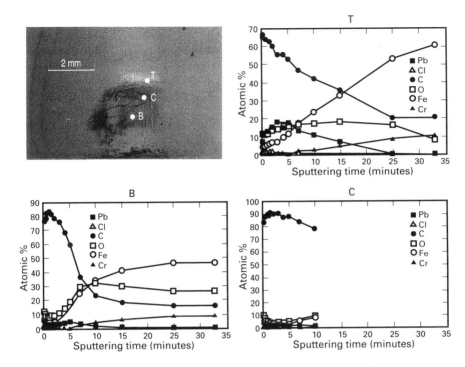

Figure 8.4   SEM micrograph (5 keV, BSE mode) and AES sputter-profile spectra of a raceway lubricant with MAC oil and Pbnp. Unlike Figures 8.1 and 8.2, this particular raceway revealed overlayers of variable Pbnp decomposition in the contact region. A film containing 15–18% Pb is present in area *T* (similar to Figures 8.1 and 8.2), while more carbonaceous films are present in areas *C* and *B*. The sputter profiling rate is estimated to be 10 nm/min. The film in area *C* is very thick, and sputtering was discontinued after 10 min

However, overlap of adjacent derivative peaks or shape changes due to variations in elemental bonding (i.e., oxidation) can distort this direct relationship, making quantitative analysis less certain in those cases. Newer AES systems avoid this problem by collecting data in summation mode.

Figure 8.4 shows profiles, with a corresponding SEM image, from a raceway lubricated with Pbnp and MAC oil as discussed in Section 8.3. The ball contact region had two types of additive films visible in the BSE mode: a bright material with AES depth profile (labeled *T*) and a dark material with AES profiles for thick and thin regions (labeled *C* and *B*, respectively). The AES data show that the bright material contains 10–20% Pb with carbon and oxygen, while the dark regions contain only 3–5% Pb with more carbon. The higher Pb composition of the film represented in *T* yields bright BSE images, while the relatively high carbon content and lower Pb content causes dark images to form in regions *C* and *B*, relative to the gray appearance of most of the steel. This result is entirely consistent with the

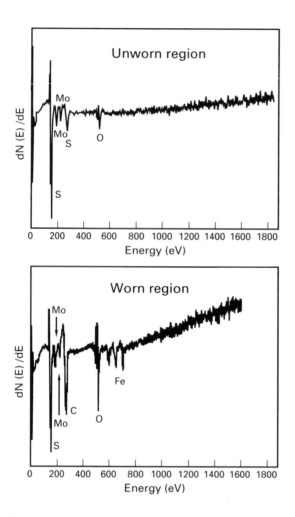

Figure 8.5 AES spectra from the raceway shown in Figure 8.3, from an unworn and worn region. The steel substrate is evident without sputtering in the worn area, clearly showing that the MoS_2 has been removed in areas. However, MoS_2 islands or debris are still present

atomic numbers of these elements as discussed in Section 8.3. These films are decomposition products of Pbnp—region T is more decomposed (i.e., has a higher Pb content) than regions C and B. It is hypothesized that region T achieved higher temperature or stress or both, for unknown reasons, than adjacent areas. Other studies have shown that these factors promote Pbnp decomposition.[10]

Figure 8.5 shows AES spectra from apparently worn and unworn regions on the MoS_2-lubricated thrust bearing shown previously in Figure 8.3. AES confirms that the worn region has exposed steel, though there are still remnant MoS_2 particles present in the irradiated regions. Spatial resolution of the particles versus the exposed regions was not possible, apparently because the beam size of the particular SAM used in this study was greater than 3 μm.

Photoelectron Spectroscopy

X-ray photoelectron spectroscopy (XPS) is used to determine the chemical composition (i.e., both the presence and bonding state of elements) present in the near-surface region (<10 nm) of a bearing. The minimum detection sensitivity of XPS is similar to AES (i.e., 1 %at. to ~3 %at.) although cross sections for particular elements are different for the two techniques. The X-ray spot size for better instruments can be <0.5 mm. Thus AES, and particularly SEM, have better spatial resolution than XPS. Obtaining high-resolution XPS data often takes longer than AES. While a single AES scan is usually accomplished in 2–5 min., only low energy resolution XPS data can be obtained in that nominal time frame. For bearing analysis, XPS is reserved for selected samples after screening by optical microscopy and SEM. (After SEM examination, representative samples are directed to either AES or XPS since destructive sputtering is normally executed in both cases to obtain depth profiles. In cases where electron beam-induced damage or contamination are likely, the investigator may choose to examine some samples with XPS immediately after optical microscopy. Comparison of these samples using XPS with other samples examined by SEM or AES prior to XPS analysis can determine if beam-induced artifacts are an issue.)

XPS of bearing surfaces can yield critical insight into the chemical mechanisms of boundary lubrication. For example, XPS has been used to examine the bearing raceways lubricated with the Pbnp/MAC discussed above. The XPS instrument used to examine the bearing raceways had a monochromatized X-ray beam with a spot size of approximately 300 μm, allowing examination both inside and outside of the contact areas. Quantitative analysis of the bearing surface composition is available with XPS. In unreacted Pbnp, XPS shows that the molecule contains approximately 3–4 %at. Pb. The much higher concentrations of Pb observed by both XPS and AES (~20%) in some bearing contact areas conclusively indicate that the Pbnp is breaking down in the contact region and that much of the hydrocarbon residue is washed away. Typically, Pb concentrations in areas outside of the contact region are closer to the 3% expected for unreacted Pbnp.

The sensitivity of XPS to chemical state is demonstrated in Figure 8.6, where the strongest peaks associated with C and Pb from Pbnp, obtained both inside and outside of a bearing contact scar, are presented. The C data show at least two principal peaks of interest, one near 285 eV associated with hydrocarbon species and one near 289 eV, indicating C bonded to the more highly electronegative O atoms in the carboxylate group of the naphthenate chain. One method of tracking the reactivity of the Pbnp is to follow the XPS intensity of the surface Pb relative to the carboxylate

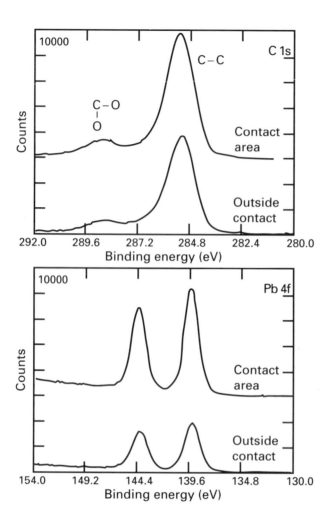

Figure 8.6 The XPS (a) C 1s peaks and (b) Pb $4f_{7/2,5/2}$ doublet obtained from a Pbnp/MAC-lubricated bearing raceway run under oscillatory conditions. The data were obtained both inside and outside of the visible contact area on the race as indicated on the diagram. The intensity ratios of the Pb $4f_{7/2}$ peak to the carboxylate C peak are approximately 6:1 inside the contact area and 3:1 outside the contact area

peak. Specifically, Figure 8.6 data obtained within a high-lead-containing wear scar show a much lower carboxylate peak intensity relative to the amount of surface lead than areas outside of the contact area. Such changes would be expected if the Pbnp was reacting with the bearing surface by losing a naphthenate chain. Ideally, chemical-state information regarding the Pb could also be obtained directly from the binding energy positions of the XPS peaks. However, the effects of the environment (specifically, oxidation of the Pb or contamination) both during testing and in sample handling prior to analysis can cloud such results. Care must be taken to ensure that environmental effects on chemical state information have been considered in any surface analysis.

The surface chemical interactions occurring in a lubricated bearing running under boundary conditions are normally so complex that a comparison to model surfaces prepared under more controlled conditions is usually beneficial. The chemistry of Pbnp on model steel surfaces has been studied with XPS and compared to the chemistry occurring in the ball-race contacts just described.[10, 11] The spectra in Figure 8.7 show the C and Pb peaks from a steel coupon immersed in a solution containing Pbnp; a region of the coupon was scratched with a diamond scribe while submerged to simulate extreme wear. The ratio of the surface Pb to the carboxylate C peak intensity in the unscratched area of the model surface was very similar to the unworn area of the bearing race, while the data obtained from the scratched area of the coupon more closely resembled the bearing contact area. The coupon experiments also followed the chemical reactions occurring on the steel surface in more detail than is possible for a real bearing. Use of coupons allowed XPS examination of peak changes occurring at high temperature in order to model the extreme conditions occurring during the boundary wear.[11] These high-temperature experiments indicate that the metallic iron present in steel bearings can chemically reduce Pbnp to metallic Pb (which can behave as a solid lubricant), a reaction which may explain the high lead concentrations found in the bearing wear scars and the boundary protection afforded by Pbnp.

The very high cross section for Pb in XPS makes XPS an extremely useful technique for the study of Pbnp reactivity on steels. In contrast, following the chemistry of the additive TCP by studying the P XPS signal is much more difficult because XPS is roughly 20 times less sensitive to P than to Pb. However, P can be observed on bearing surfaces treated with TCP or run with TCP containing lubricants. Although much more analysis time is needed to obtain chemical state information than for Pb, the XPS data indicate that the P remains in a phosphate-like bonding moiety upon adsorption to steel. Possible chemical reactions that might occur during bearing operation are currently being studied. Generally, it is extremely important to understand the strengths and limitations of XPS in studying bearing surface chemistry before embarking on such an experiment and during analyses of the data obtained in such an experiment.

XPS has been used very little, to date, to study sputter-deposited MoS_2 films on actual bearings due to the limited spatial resolution of the technique and the difficulty of examining curved surfaces. XPS has been used quite successfully in examining witness plates used in deposition runs of MoS_2. A recent investigation studied the oxygen in these films, stemming from the deposition process and from post-deposition sources.[12] In particular, phases of $MoS_{2-x}O_x$ containing substitutional oxygen were identified.

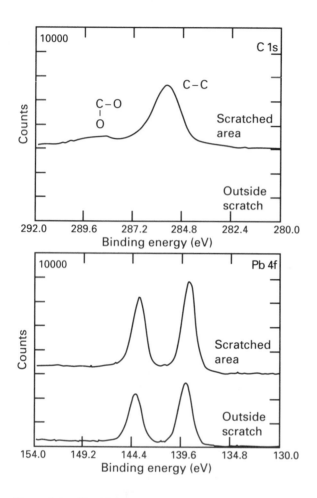

Figure 8.7 The XPS (a) C 1s peaks and (b) Pb 4f$_{7/2,5/2}$ doublet from a Pbnp-treated 440C stainless steel surface. The data were obtained from an undisturbed portion of the surface and from an area scratched during the Pbnp immersion treatment. The Pb:carboxylate peak ratios in the scratched areas are very similar to those observed in the contact region of the bearing in Figure 8.6

SIMS

An alternative to AES in obtaining elemental information is to use ion-scattering techniques such as secondary ion mass spectrometry (SIMS)[13] or secondary neutral mass spectrometry (SNMS).[14] The major advantage of these techniques is better sensitivity: the sensitivity of SIMS ranges from 10^{-6} to 0.1 %at., depending upon the element; SNMS is generally 10^{-3} %at.; while that of AES is 0.1–1 %at. The spatial resolution of SIMS/SNMS is not as good as AES. These techniques are inherently destructive since sputtering of the surface is required for data collection.

Vibrational Spectroscopy

Vibrational spectroscopies can be used to obtain information about organic residues on bearing surfaces. In particular, Fourier-transform infrared (FTIR) microscopes can provide spatially resolved structural information on overlayer films. The technique does not, however, have the surface sensitivity of AES, XPS, or SIMS/SNMS. High-resolution electron energy loss spectroscopy (HREELS) does provide surface-sensitive vibrational information, which is particularly useful in elucidating the structure of thin organic molecular layers. However, HREELS does not appear to be in use for bearing analysis at this time. Investigators using this technique seem to be using small samples that provide model surfaces for surface science research.

8.5 Future Directions

Improved equipment for established techniques and new techniques are continually being introduced for the surface analysis of bearings. High spatial resolution SAMs are becoming available that are effectively UHV SEMs as well as SAMs. Scanning probe microscopies, such as scanning tunneling microscopy (STM) and atomic force microscopy (AFM), are becoming widely used in tribology research. STM/AFM have not been used significantly in bearing analysis. These microscopies can provide, at a minimum, atomic resolution structural information and profilometry, and their use should increase to complement conventional profilometry. It would be very beneficial and convenient for bearing analysis to have a STM or AFM integrated into a SEM or in a high spatial resolution SAM. The SEM/SAM could view the bearing over a wide range of magnifications, while the STM/AFM could be dropped onto selected small regions for detailed investigation.

The use of XPS to study bearing surfaces has been limited by several factors, including X-ray spot size, fairly lengthy data accumulation periods, and inflexibility in sample handling procedures. Improvements are continually being made in generating smaller diameter X-ray beams that are still capable of providing a sufficient signal-to-noise ratio for data collection. These smaller analysis areas should allow the study of smaller wear features in bearings. In addition, techniques such as photoelectron microscopy are currently being developed that should allow detailed chemical mapping of surfaces of interest. The lengthy data collection times have been considerably shortened with multichannel detectors and high-efficiency electron optics, but low-concentration species with very low cross sections still present difficulties for analysis. The introduction of new X-ray anode materials or even synchrotron radiation providing photons with lower energies to enhance the sensitivity of the technique to particular bearing surface species is certainly possible. Finally, most surface science instruments are not designed to handle specimens having curved shapes such as bearing raceways. It is likely that custom sample handling mechanisms will have to be developed for particular applications.

Nonlinear optical spectroscopies, such as second-harmonic generation and sum-frequency generation, are under development.[14] These techniques could provide surface and interfacial characterization at ambient pressures, avoiding the vacuum environments required for the electron spectroscopies reviewed in this text.

Finally, semiconductor-based microsensors could be used to monitor the health of a mechanism during operation. A microaccelerometer package is currently flying on the Olympus spacecraft.[16] The accelerometers measure the vibration modes induced by the operation of mechanisms on the spacecraft. In the future, mass spectrometry could be used to investigate the gas-phase by-products of lubricant degradation that may occur during bearing tests of space applications. If such studies established gas-phase signatures to lubricant degradation, semiconductor-based chemical sensors could be tailored to monitor these ejected molecules. These sensors could be incorporated into the bearing mechanism as part of a health monitoring scheme to notify the user or an automated controller that can respond that additional lubrication is needed. The dawn of the "smart bearing" may be near.

Acknowledgments

This work was supported by the Air Force Materiel Command, Space and Missile Systems Center, under Contract F04701-88-C-0089. The authors wish to thank D. J. Carré, P. D. Fleischauer, and J. R. Lince at Aerospace and G. Hopple (Lockheed Missiles & Space Co.) for many helpful discussions over the years.

References

1 E.N. Bamberger, *Tribology in the 80s*, NASA Publication 2300, Vol. II, 774, 1985.

2 E. Rabinowicz, *Friction and Wear of Materials*, Wiley, New York, 1965.

3 G. Hopple, Failure analysis of precision bearings, in *Proc. of the Advanced Materials Symposium*, ISTFA '87, ASM International 1987, 201–208.

4 Y.-W. Chung, A.M. Homola, and G.B. Street, Surface science investigations in tribology—experimental approaches, *ACS Symposium Series 485*, 1992.

5 D.J. Carré, R. Bauer, and P.D. Fleischauer, *ASLE Transactions*, **26** (4), 475–480.

6 T.E. Tallian, *Failure Atlas for Hertz Contact Machine Elements*, ASME Press, Fairfield, 1992.

7 L. Engel and H. Kingele, *An Atlas of Metal Damage*, Prentice-Hall, Englewood Cliffs, 1974.

8 J.G. Barbee, and F.M. Newman, Thickener/fluid interaction in lubricating greases, Southwest Research Inst., San Antonio; US Army Belvoir Research and Development Center, AFLRL Report No. 173, April 1984, Gov. Accession No. AD-A142-276.

9 P.D. Fleischauer and R. Bauer, *Tribology Trans.*, **31** (2), 239, 1988.

10 S.V. Didziulis, M.R. Hilton, and P.D. Fleischauer, The influence of steel surface chemistry on the bonding of lubricant films, in *ACS Symp. Series* **485**, 1992, 43–57.

11 S.V. Didziulis and P.D. Fleischauer, *Langmuir*, 7, 2981, 1991.

12 J. R. Lince, M.R. Hilton, and A.S. Bommannavar, *Surface and Coatings Technol.*, **43/44**, 640, 1990.

13 P. Williams, *Ann. Rev. Mater. Sci.*, **15**, 517, 1985.

14 K.-H. Muller, K. Seifert, and M. Wilmers, *J. Vac. Sci. Technol.*, **A3**(3), 1367, 1985.

15 Y.R. Shen, *Ann. Rev. Mater. Sci.*, **16**, 69, 1986.

16 D. Turnbridge, The PAX experiment on Olympus, *ESA Bulletin*, **64**, 89 1990.

Appendixes: Technique Summaries

The technique summaries marked with an asterisk in the following pages are reprinted from the lead volume of this series, *Encyclopedia of Materials Characterization*, by C. Richard Brundle, Charles A. Evans, Jr., and Shaun Wilson; they are summaries of full-length articles appearing there.

Light Microscopy* 1

The light microscope uses the visible or near visible portion of the electromagnetic spectrum; light microscopy is the interpretive use of the light microscope. This technique, which is much older than other characterization instruments, can trace its origin to the 17th century. Modern analytical and characterization methods began about 150 years ago when thin sections of rocks and minerals, and the first polished metal and metal-alloy specimens were prepared and viewed with the intention of correlating their structures with their properties. The technique involves, at its very basic level, the simple, direct visual observation of a sample with white-light resolution to 0.2 μm. The morphology, color, opacity, and optical properties are often sufficient to characterize and identify a material.

Range of samples characterized	Almost unlimited for solids and liquid crystals
Destructive	Usually nondestructive; sample preparation may involve material removal
Quantification	Via calibrated eyepiece micrometers and image analysis
Detection limits	To sub-ng
Resolving power	0.2 μm with white light
Imaging capabilities	Yes
Main use	Direct visual observation; preliminary observation for final characterization, or preparative for other instrumentation
Instrument cost	$2,500–$50,000 or more
Size	Pocket to large table

Scanning Electron Microscopy (SEM)*

The scanning electron microscope (SEM) is often the first analytical instrument used when a "quick look" at a material is required and the light microscope no longer provides adequate resolution. In the SEM an electron beam is focused into a fine probe and subsequently raster scanned over a small rectangular area. As the beam interacts with the sample it creates various signals (secondary electrons, internal currents, photon emission, etc.), all of which can be appropriately detected. These signals are highly localized to the area directly under the beam. By using these signals to modulate the brightness of a cathode ray tube, which is raster scanned in synchronism with the electron beam, an image is formed on the screen. This image is highly magnified and usually has the "look" of a traditional microscopic image but with a much greater depth of field. With ancillary detectors, the instrument is capable of elemental analysis.

Main use	High magnification imaging and composition (elemental) mapping
Destructive	No, some electron beam damage
Magnification range	10×–300,000×; 5000×–100,000× is the typical operating range
Beam energy range	500 eV–50 keV; typically, 20–30 keV
Sample requirements	Minimal, occasionally must be coated with a conducting film; must be vacuum compatible
Sample size	Less than 0.1mm, up to 10 cm or more
Lateral resolution	1–50 nm in secondary electron mode
Depth sampled	Varies from a few nm to a few μm, depending upon the accelerating voltage and the mode of analysis
Bonding information	No
Depth profiling capabilities	Only indirect
Instrument cost	$100,000–$300,000 is typical
Size	Electronics console 3 ft. × 5 ft.; electron beam column 3 ft. × 3 ft.

In Situ Wear Device for the Scanning Electron Microscope 3

A large chamber SEM can be adapted to perform wear experiments so that wear processes can be observed in real-time and at high magnification. The usual configuration is a pin-on-disk apparatus. The pin is usually needle shaped so that the contact area can be readily observed. The SEM should be able to operate in the TV mode if dynamic wear processes are to be observed. When observing in the TV mode, the rotation of the disk should be slowed to prevent blurring of the image. The higher the magnification used, the lower the speed of rotation. A maximum magnification of 800× to 900× is recommended.

The disk can be mounted on the SEM stage and rotated by the system used to rotate the stage. The pin assembly can then be mounted on a stationary part of the stage and the pin spring loaded. The wear apparatus can also be self contained with an electric drive motor and the assembly designed to mount on the stage. Temperature and friction force measuring transducers can be included in the wear device and the electric leads brought out through any contact probes which penetrate the SEM chamber

If the electron source is separated from the main chamber by a diaphragm with a pin hole for the electron beam, a high vacuum can be obtained around the electron source and the stage area can operate at low vacuum for experiments with organic coatings or liquid lubricating films.

Range of elements	Those with electrical conductivity
Destructive	No
Magnification range	800–900× best maximum for real-time dynamic observations
Sample size	Disk diameter 1in.–3 in. pin tip radius 20–100 μm for dynamic wear observation
Lateral resolution	1–50 nm
Instrument cost	$100,000–$300,000
Size	Chamber 2 ft. × 2 ft., beam column 3 ft. × 3 ft., console 3 ft. × 5 ft.

Scanning Tunneling Microscopy and Scanning Force Microscopy (STM and SFM)* 4

In scanning tunneling microscopy (STM) or scanning force microscopy (SFM), a solid specimen in air, liquid or vacuum is scanned by a sharp tip located within a few Å of the surface. In STM, a quantum-mechanical tunneling current flows between atoms on the surface and those on the tip. In SFM, also known as Atomic Force Microscopy (AFM), interatomic forces between the atoms on the surface and those on the tip cause the deflection of a microfabricated cantilever. Because the magnitude of the tunneling current or cantilever deflection depends strongly upon the separation between the surface and tip atoms, they can be used to map out surface topography with atomic resolution in all three dimensions. The tunneling current in STM is also a function of local electronic structure so that atomic-scale spectroscopy is possible. Both STM and SFM are unsurpassed as high-resolution, three-dimensional profilometers.

Parameters measured	Surface topography (SFM and STM); local electronic structure (STM)
Destructive	No
Vertical resolution	STM, 0.01 Å; SFM, 0.1 Å
Lateral resolution	STM, atomic; SFM, atomic to 1 nm
Quantification	Yes; three-dimensional
Accuracy	Better than 10% in distance
Imaging/mapping	Yes
Field of view	From atoms to > 250 μm
Sample requirements	STM—solid conductors and semiconductors, conductive coating required for insulators; SFM—solid conductors, semiconductors and insulators
Main uses	Real-space three-dimensional imaging in air, vacuum, or solution with unsurpassed resolution; high-resolution profilometry; imaging of nonconductors (SFM)
Instrument cost	$65,000 (ambient) to $200,000 (ultrahigh vacuum)
Size	Table-top (ambient), 2.27–12 inch bolt-on flange (ultrahigh vacuum)

Transmission Electron Microscopy (TEM)*   5

In transmission electron microscopy (TEM) a thin solid specimen ($\leq$ 200 nm thick) is bombarded in vacuum with a highly-focused, monoenergetic beam of electrons. The beam is of sufficient energy to propagate through the specimen. A series of electromagnetic lenses then magnifies this transmitted electron signal. Diffracted electrons are observed in the form of a diffraction pattern beneath the specimen. This information is used to determine the atomic structure of the material in the sample. Transmitted electrons form images from small regions of sample that contain contrast, due to several scattering mechanisms associated with interactions between electrons and the atomic constituents of the sample. Analysis of transmitted electron images yields information both about atomic structure and about defects present in the material.

Range of elements	TEM does not specifically identify elements measured
Destructive	Yes, during specimen preparation
Chemical bonding information	Sometimes, indirectly from diffraction and image simulation
Quantification	Yes, atomic structures by diffraction; defect characterization by systematic image analysis
Accuracy	Lattice parameters to four significant figures using convergent beam diffraction
Detection limits	One monolayer for relatively high-Z materials
Depth resolution	None, except there are techniques that measure sample thickness
Lateral resolution	Better than 0.2 nm on some instruments
Imaging/mapping	Yes
Sample requirements	Solid conductors and coated insulators. Typically 3-mm diameter, < 200-nm thick in the center
Main uses	Atomic structure and Microstructural analysis of solid materials, providing high lateral resolution
Instrument cost	\$300,000–\$1,500,000
Size	100 ft.^{2} to a major lab

Energy-Dispersive X-Ray Spectroscopy (EDS)*   6

When the atoms in a material are ionized by a high-energy radiation they emit characteristic X rays. EDS is an acronym describing a technique of X-ray spectroscopy
that is based on the collection and energy dispersion of characteristic X rays. An
EDS system consists of a source of high-energy radiation, usually electrons; a sample; a solid state detector, usually made from lithium-drifted silicon, Si (Li); and
signal processing electronics. EDS spectrometers are most frequently attached to
electron column instruments. X rays that enter the Si (Li) detector are converted
into signals which can be processed by the electronics into an X-ray energy histogram. This X-ray spectrum consists of a series of peaks representative of the type
and relative amount of each element in the sample. The number of counts in each
peak may be further converted into elemental weight concentration either by comparison with standards or by standardless calculations.

Range of elements	Boron to uranium
Destructive	No
Chemical bonding information	Not readily available
Quantification	Best with standards, although standardless methods are widely used
Accuracy	Nominally 4–5%, relative, for concentrations > 5% wt.
Detection limits	100–200 ppm for isolated peaks in elements with $Z > 11$, 1–2% wt. for low-Z and overlapped peaks
Lateral resolution	0.5–1 μm for bulk samples; as small as 1 nm for thin samples in STEM
Depth sampled	0.02 to μm, depending on Z and keV
Imaging/mapping	In SEM, EPMA, and STEM
Sample requirements	Solids, powders, and composites; size limited only by the stage in SEM, EPMA and XRF; liquids in XRF; 3 mm diameter thin foils in TEM
Main use	To add analytical capability to SEM, EPMA and TEM
Cost	$25,000–$100,000, depending on accessories (not including the electron microscope)

Scanning Transmission Electron Microscopy (STEM)* 7

In scanning transmission electron microscopy (STEM) a solid specimen, 5–500 nm thick, is bombarded in vacuum by a beam (0.3–50 nm in diameter) of monoenergetic electrons. STEM images are formed by scanning this beam in a raster across the specimen and collecting the transmitted or scattered electrons. Compared to the TEM an advantage of the STEM is that many signals may be collected simultaneously: bright- and dark-field images; convergent beam electron diffraction (CBED) patterns for structure analysis; and energy-dispersive X-ray spectrometry (EDS) and electron energy-loss spectrometry (EELS) signals for compositional analysis. Taken together, these analysis techniques are termed analytical electron microscopy (AEM). STEM provides about 100 times better spatial resolution of analysis than conventional TEM. When electrons scattered into high angles are collected, extremely high-resolution images of atomic planes and even individual heavy atoms may be obtained.

Range of elements	Lithium to uranium
Destructive	Yes, during specimen preparation
Chemical bonding information	Sometimes, from EELS
Quantification	Quantitative compositional analysis from EDS or EELS, and crystal structure analysis from CBED
Accuracy	5–10% relative for EDS and EELS
Detection limits	0.1–3.0% wt. for EDS and EELS
Lateral resolution	Imaging, 0.2–10 nm; EELS, 0.5–10 nm; EDS, 3–30 nm
Imaging/mapping capabilities	Yes, lateral resolution down to < 5 nm
Sample requirements	Solid conductors and coated insulators typically 3 mm in diameter and < 200 nm thick at the analysis point for imaging and EDS, but < 50 nm thick for EELS
Main uses	Microstructural, crystallographic, and compositional analysis; high spatial resolution with good elemental detection and accuracy; unique structural analysis with CBED
Instrument cost	$500,000–$2,000,000
Size	3 m × 4 m × 3 m

Electron Probe X-Ray Microanalysis (EPMA)* 8

Electron probe X-ray microanalysis (EPMA) is an elemental analysis technique based upon bombarding a specimen with a focused beam of energetic electrons (beam energy 5–30 keV) to induce emission of characteristic X rays (energy range 0.1–15 keV). The X rays are measured by energy-dispersive (EDS) or wavelength-dispersive (WDS) X-ray spectrometers. Quantitative matrix (interelement) correction procedures based upon first principles physical models provide great flexibility in attacking unknown samples of arbitrary composition; the standards suite can be as simple as pure elements or binary compounds. Typical error distributions are such that relative concentration errors lie within ±4% for 95% of cases when the analysis is performed with pure element standards. Spatial distributions of elemental constituents can be visualized qualitatively by X-ray area scans (dot maps) and quantitatively by digital compositional maps.

Range of elements	Beryllium to the actinides
Destructive	No, except for electron beam damage
Chemical bonding	In rare cases: from light-element X-ray peak shifts
Depth profiling	Rarely, by changing incident beam energy
Quantification	Standardless or; pure element standards
Accuracy	±4% relative in 95% of cases; flat, polished samples
Detection limits	WDS, 100 ppm; EDS, 1000 ppm
Sampling depth	Energy and matrix dependent, 100 nm–5 μm
Lateral resolution	Energy and matrix dependent, 100 nm–5 μm
Imaging/mapping	Yes, compositional mapping and SEM imaging
Sample requirements	Solid conductors and insulators; typically, < 2.5 cm in diameter, and < 1 cm thick, polished flat; particles, rough surfaces, and thin films
Major uses	Accurate, nondestructive quantitative analysis of major, minor, and trace constituents of materials
Instrument cost	$300,000–$800,000
Size	3 m × 1.5 m × 2 m high

X = ?c   X = d sin θ

X-Ray Diffraction (XRD)* 9

In X-ray diffraction (XRD) a collimated beam of X rays, with wavelength $\lambda \sim 0.5$–2 Å, is incident on a specimen and is diffracted by the crystalline phases in the specimen according to Bragg's law ($\lambda = 2d \sin\theta$, where d is the spacing between atomic planes in the crystalline phase). The intensity of the diffracted X rays is measured as a function of the diffraction angle 2θ and the specimen's orientation. This diffraction pattern is used to identify the specimen's crystalline phases and to measure its structural properties, including strain (which is measured with great accuracy), epitaxy, and the size and orientation of crystallites (small crystalline regions). XRD can also determine concentration profiles, film thicknesses, and atomic arrangements in amorphous materials and multilayers. It also can characterize defects. To obtain this structural and physical information from thin films, XRD instruments and techniques are designed to maximize the diffracted X-ray intensities, since the diffracting power of thin films is small.

Range of elements	All, but not element specific. Low-Z elements may be difficult to detect
Probing depth	Typically a few μm, but material dependent; monolayer sensitivity with synchrotron radiation
Detection Limits	Material dependent, but ~3% in a two phase mixture; with synchrotron radiation can be ~0.1%
Destructive	No, for most materials
Depth profiling	Normally no; but this can be achieved.
Sample requirements	Any material, greater than ~0.5 cm, although smaller with microfocus
Lateral resolution	Normally none; although ~10 μm with microfocus
Main use	Identification of crystalline phases; determination of strain, and crystallite orientation and size; accurate determination of atomic arrangements
Specialized uses	Defect imaging and characterization; atomic arrangements in amorphous materials and multilayers; concentration profiles with depth; film thickness measurements
Instrument cost	$70,000–$200,000
Size	Varies with instrument, greater than ~70 ft.^{2}

Low-Energy Electron Diffraction (LEED)* 10

In low-energy electron diffraction (LEED) a collimated monoenergetic beam of electrons in the energy range 10–1000 eV ($\lambda \approx 0.4$–4.0 Å) is diffracted by a specimen surface. In this energy range, the mean free path of electrons is only a few Å, leading to surface sensitivity. The diffraction pattern can be analyzed for the existence of a clean surface or an ordered overlayer structure. Intensities of diffracted beams can be analyzed to determine the positions of surface atoms relative to each other and to underlying layers. The shapes of diffracted beams in angle can be analyzed to provide information about surface disorder. Various phenomena related to surface crystallography and microstructure can be investigated. This technique requires a vacuum.

Range of elements	All elements, but not element specific
Destructive	No, except in special cases of electron-beam damage
Depth probed	4–20 Å
Detection limits	0.1 monolayer; any ordered phase can be detected; atomic positions to 0.1 Å; step heights to 0.1 Å; surface disorder down to ~10% of surface sites
Resolving power	Maximum resolvable distance for detecting disorder: typically 200 Å; best systems, 5 µm
Lateral resolution	Typical beam sizes, 0.1 mm; best systems, ~10 µm
Imaging capability	Typically, no; with specialized instruments (e.g., low-energy electron microscopy), 150 Å
Sample requirements	Single crystals of conductors and semiconductors; insulators and polycrystalline samples under special circumstances; 0.25 cm^2 or larger, smaller with special effort
Main uses	Analysis of surface crystallography and microstructure; surface cleanliness
Cost	≤$75,000; can be home built cheaply
Size	Generally part of other systems; if self-standing, ~8 m^2

X-Ray Photoelectron Spectroscopy (XPS)* 11

In X-ray photoelectron spectroscopy (XPS) monoenergetic soft X rays bombard a sample material, causing electrons to be ejected. Identification of the elements present in the sample can be made directly from the kinetic energies of these ejected photoelectrons. On a finer scale it is also possible to identify the chemical state of the elements present from small variations in the determined kinetic energies. The relative concentrations of elements can be determined from the measured photoelectron intensities. For a solid, XPS probes 2–20 atomic layers deep, depending on the material, the energy of the photoelectron concerned, and the angle (with respect to the surface) of the measurement. The particular strengths of XPS are semiquantitative elemental analysis of surfaces without standards, and chemical state analysis, for materials as diverse as biological to metallurgical. XPS also is known as electron spectroscopy for chemical analysis (ESCA).

Range of elements	All except hydrogen and helium
Destructive	No, some beam damage to X-ray sensitive materials
Elemental analysis	Yes, semiquantitative without standards; quantitative with standards. Not a trace element method
Chemical state information	Yes
Depth probed	5–50 Å
Depth profiling	Yes, over the top 50 Å; greater depths require sputter profiling
Depth resolution	A few to several tens of Å, depending on conditions
Lateral resolution	5 mm to 75 μm; down to 5 μm in special instruments
Sample requirements	All vacuum-compatible materials; flat samples best; size accepted depends on particular instrument
Main uses	Determinations of elemental and chemical state compositions in the top 30 Å
Instrument cost	$200,000–$1,000,000, depending on capabilities
Size	10 ft. × 12 ft.

Auger Electron Spectroscopy (AES)* 12

Auger electron spectroscopy (AES) uses a focused electron beam to create secondary electrons near the surface of a solid sample. Some of these (the Auger electrons) have energies characteristic of the elements, and in many cases, of the chemical bonding of the atoms from which they are released. Because of their characteristic energies and the shallow depth from which they escape without energy loss, Auger electrons are able to characterize the elemental composition and, at times, the chemistry of the surfaces of samples. When used in combination with ion sputtering to gradually remove the surface, Auger spectroscopy can similarly characterize the sample in depth. The high spatial resolution of the electron beam and the process allows microanalysis of three-dimensional regions of solid samples. AES has the attributes of high lateral resolution, relatively high sensitivity, standardless semiquantitative analysis, and chemical bonding information in some cases.

Range of elements	All except H and He
Destructive	No, except to electron beam-sensitive materials and during depth profiling
Elemental Analysis	Yes, semiquantitative without standards; quantitative with standards
Absolute sensitivity	100 ppm for most elements, depending on the matrix
Chemical state information?	Yes, in many materials
Depth probed	5–100 Å
Depth profiling	Yes, in combination with ion-beam sputtering
Lateral resolution	300 Å for Auger analysis, even less for imaging
Imaging/mapping	Yes, called Scanning Auger Microscopy, SAM
Sample requirements	Vacuum-compatible materials
Main use	Elemental composition of inorganic materials
Instrument cost	$100,000–$800,000
Size	10 ft. × 15 ft.

Fourier Transform Infrared Spectroscopy (FTIR)* 13

The vibrational motions of the chemically bound constituents of matter have frequencies in the infrared regime. The oscillations induced by certain vibrational modes provide a means for matter to couple with an impinging beam of infrared electromagnetic radiation and to exchange energy with it when the frequencies are in resonance. In the infrared experiment, the intensity of a beam of infrared radiation is measured before (I_0) and after (I) it interacts with the sample as a function of light frequency (w_i). A plot of I/I_0 versus frequency is the "infrared spectrum." The identities, surrounding environments, and concentrations of the chemical bonds that are present can be determined.

Information	Vibrational frequencies of chemical bonds
Element Range	All, but not element specific
Destructive	No
Chemical bonding information	Yes, identification of functional groups
Depth profiling	No, not under standard conditions
Depth Probed	Sample dependent, from 1 µm to 10 nm
Detection limits	Ranges from undetectable to $< 10^{13}$ bonds/cc. Sub-monolayer sometimes
Quantification	Standards usually needed
Reproducibility	0.1% variation over months
Lateral resolution	0.5 cm to 20 µm
Imaging/mapping	Available, but not routinely used
Sample requirements	Solid, liquid, or gas in all forms; vacuum not required
Main use	Qualitative and quantitative determination of chemical species, both trace and bulk, for solids and thin films. Stress, structural inhomogeneity
Instrument cost	$50,000–$150,000 for FTIR; $20,000 or more for non-FT spectrophotometers
Instrument size	Ranges from desktop to 2 m × 2 m

Raman Spectroscopy* **14**

Raman spectroscopy is the measurement, as a function of wavenumber, of the inelastic light scattering that results from the excitation of vibrations in molecular and crystalline materials. The excitation source is a single line of a continuous gas laser, which permits optical microscope optics to be used for measurement of samples down to a few μm. Raman spectroscopy is sensitive to molecular and crystal structure; applications include chemical fingerprinting, examination of single grains in ceramics and rocks, single-crystal measurements, speciation of aqueous solutions, identification of compounds in bubbles and fluid inclusions, investigations of structure and strain states in polycrystalline ceramics, glasses, fibers, gels, and thin and thick films.

Information	Vibrational frequencies of chemical bonds
Element range	All, but not element specific
Destructive	No, unless sample is susceptible to laser damage
Lateral resolution	1 μm with microfocus instruments
Depth profiling	Limited to transparent materials
Depth probed	Few μm to mm, depending on material
Detection limits	1000 Å normally, submonolayer in special cases
Quantitative	With difficulty; usually qualitative only
Imaging	Usually no, although imaging instruments have been built
Sample requirements	Very flexible: liquids, gases, crystals, polycrystalline solids, powders, and thin films
Main use	Identification of unknown compounds in solutions, liquids, and crystalline materials; characterization of structural order, and phase transitions
Instrument cost	$150,000–$250,000
Size	1.5 m × 2.5 m

Rutherford Backscattering Spectrometry (RBS)* 15

Rutherford backscattering spectrometry (RBS) analysis is performed by bombarding a sample target with a monoenergetic beam of high-energy particles, typically helium, with an energy of a few MeV. A fraction of the incident atoms scatter backwards from heavier atoms in the near-surface region of the target material, and usually are detected with a solid state detector that measures their energy. The energy of a backscattered particle is related to the depth and mass of the target atom, while the number of backscattered particles detected from any given element is proportional to concentration. This relationship is used to generate a quantitative depth profile of the upper 1–2 μm of the sample. Alignment of the ion beam with the crystallographic axes of a sample permits crystal damage and lattice locations of impurities to be quantitatively measured and depth profiled. The primary applications of RBS are the quantitative depth profiling of thin-film structures, crystallinity, dopants, and impurities.

Range of elements	Lithium to uranium
Destructive	$\sim10^{13}$ He atoms implanted; radiation damage
Chemical bonding information	No
Quantification	Yes, standardless; accuracy 5-20%
Detection limits	10^{12}–10^{16} atoms/cm^2; 1–10 at.% for low-Z elements; 0–100 ppm for high-Z elements
Lateral resolution	1–4 mm, 1 μm in specialized equipment
Depth profiling	Yes, and nondestructive
Depth resolution	2–30 nm
Maximum depth	$\sim$2 μm, 20 μm with H$^+$
Imaging/mapping	Under development
Sample requirements	Solid, vacuum compatible
Main use	Nondestructive depth profiling of thin films, crystal damage information
Instrument cost	$450,000–$1,000,000
Size	2 m × 7 m

Static Secondary Ion Mass Spectrometry (Static SIMS)* 16

Static secondary ion mass spectrometry (SIMS) involves the bombardment of a sample with an energetic (typically 1–10 keV) beam of particles, which may be either ions or neutrals. As a result of the interaction of these primary particles with the sample, species are ejected that have become ionized. These ejected species, known as secondary ions, are the analytical signal in SIMS.

In static SIMS, the use of a low dose of incident particles (typically less than 5×10^{12} atoms/cm^2) is critical to maintain the chemical integrity of the sample surface during analysis. A mass spectrometer sorts the secondary ions with respect to their specific charge-to-mass ratio, thereby providing a mass spectrum composed of fragment ions of the various functional groups or compounds on the sample surface. The interpretation of these characteristic fragmentation patterns results in a chemical analysis of the outer few monolayers. The ability to obtain surface chemical information is the key feature distinguishing static SIMS from dynamic SIMS, which profiles rapidly into the sample, destroying the chemical integrity of the sample.

Range of elements	H to U; all isotopes
Destructive	Yes, if sputtered long enough
Chemical bonding information	Yes
Depth probed	Outer 1 or 2 monolayers
Lateral resolution	Down to ~100 μm
Imaging/mapping	Yes
Quantification	Possible with appropriate standards
Mass range	Typically, up to 1000 amu (quadrupole), or up to 10,000 amu (time of flight)
Sample requirements	Solids, liquids (dispersed or evaporated on a substrate), or powders; must be vacuum compatible
Main use	Surface chemical analysis, particularly organics, polymers
Instrument cost	$500,000–$750,000
Size	4 ft. × 8 ft.

Surface Roughness: Measurement, Formation by Sputtering, Impact on Depth Profiling* 17

Surface roughness is commonly measured using mechanical and optical profilers, scanning electron microscopes, and atomic force and scanning tunneling microscopes. Angle-resolved scatterometers can also be applied to this measurement. The analysis surface can be roughened by ion bombardment, and roughness will degrade depth resolution in a depth profile. Rotation of the sample during sputtering can reduce this roughening.

Mechanical Profiler

Depth resolution	0.5 nm
Minimum step	2.5–5 nm
Maximum step	~150 μm
Lateral resolution	0.1–25 μm, depending on stylus radius
Maximum sample size	15-mm thickness, 200-mm diameter
Instrument cost	$30,000–$70,000

Optical Profiler

Depth resolution	0.1 nm
Minimum step	0.3 nm
Maximum step	15 μm
Lateral resolution	0.35–9 μm, depending on optical system
Maximum sample size	125-mm thickness, 100-mm diameter
Instrument cost	$80,000–$100,000

SEM (see SEM article)

Scanning Force Microscope (see STM/SFM article)

Depth resolution	0.01 nm
Lateral resolution	0.1 nm
Instrument cost	$75,000–$150,000

Scanning Tunneling Microscope (see STM/SFM article)

Depth resolution	0.001 μm
Lateral resolution	0.1 nm
Instrument cost	$75,000–$150,000

Optical Scatterometer

Depth resolution	0.1 nm (root mean square)
Instrument cost	$50,000–$150,000

Index